Utilization of Pectin in the Food and Drug Industries

Edited by Maruf Ahmed

Published in London, United Kingdom

Utilization of Pectin in the Food and Drug Industries
http://dx.doi.org/10.5772/intechopen.105289
Edited by Maruf Ahmed

Contributors
Haneef Ur Rehman, Wasiu Awoyale, Kazeem K. Olatoye, Busie Maziya-Dixon, Joan Oñate Narciso, Nils Rentsch, Laura Nyström, César Pérez-Alonso, Stefani Cortés-Camargo, Angélica Román-Guerrero, Erik Alpizar-Reyes, Roshanak Vakili, Chuckwuemeka K. Nkere, Nneka R. Rita Okereke

First published in London, United Kingdom, 2023 by IntechOpen
IntechOpen is the global imprint of INTECHOPEN LIMITED, registered in England and Wales, registration number: 11086078, 5 Princes Gate Court, London, SW7 2QJ, United Kingdom

British Library Cataloguing-in-Publication Data
A catalogue record for this book is available from the British Library

Additional hard and PDF copies can be obtained from orders@intechopen.com

Utilization of Pectin in the Food and Drug Industries
Edited by Maruf Ahmed
p. cm.
Print ISBN 978-1-83768-493-9
Online ISBN 978-1-83768-494-6
eBook (PDF) ISBN 978-1-83768-495-3

Meet the editor

Dr. Maruf Ahmed is a professor in the Department of Food Processing and Preservation at Hajee Mohammad Danesh Science and Technology University in Dinajpur, Bangladesh. From 2018 to 2020 he was Dean of the Faculty of Engineering at the same institution. Maruf obtained his doctorate in food science and technology from Chonnam National University in Gwangju, Republic of Korea. His areas of specialization include food and nutrition security, food value addition, agricultural value chain development, product development, antioxidants, and biodegradable plastic. Dr. Ahmed has over 50 publications in peer-reviewed journals, as well as a number of book chapters.

Contents

Preface

Scientists and technologists are particularly interested in pectin for its nutritional content and because of its versatile uses since ancient times as an important source of food fiber and as a gelling agent in foods. This book is a cutting-edge, multidisciplinary reference work that will promote pectin research by combining and fostering new perspectives and methodologies from different fields of study. The extraction, processing, and uses of pectin in the food and pharmaceutical industries are all covered.

I'd like to express my deepest gratitude to all the authors who have worked on this publication. Last but not least, I'd like to thank my family and other supporters who have helped me stay motivated during this project.

Dr. Maruf Ahmed
Professor,
Department of Food Processing and Preservation,
Hajee Mohammad Danesh Science and Technology University,
Dinajpur, Bangladesh

Chapter 1

Extraction of Pectin from Orange Peel Wastes as an Ingredient for Edible Films Containing Kabog Millet Flour

Nils Rentsch, Laura Nyström and Joan Oñate Narciso

Abstract

Fossil-based plastic is a popular material for food packaging. It can cause negative environmental consequences due to its low biodegradability. To address this challenge, a pectin-based edible plastic with added nutritional value by incorporating whole-grain kabog millet flour was prepared. The pectin in the films was extracted by microwave-assisted and enzymatic procedures from orange peel wastes. The extracted pectin was tested for its degree of esterification using Fourier-transform infrared spectroscopy, its molecular weight and behavior in aqueous solutions using size-exclusion chromatography, and its monosaccharide composition using ion-exchange chromatography. Biodegradable and edible pectin films were produced and tested for their mechanical properties: maximum strain, maximum stress, and water contact angle. The results showed a significant increase in hydrophobicity of the film surface by adding whole-grain kabog millet flour. The maximum strain of the film, however, was reduced to around 80% upon the addition of the whole-grain kabog millet flour. Enzymatically-extracted pectin increased the film hydrophobicity. Hydrophobic surfaces have higher water resistance; thus, the enzymatically-extracted pectin can be developed for further applications. Due to the low elasticity of the films, a possible application would be as direct coating of fruits and vegetables incorporating antioxidants or antimicrobials.

Keywords: pectin, microwave-assisted, enzymatic extraction, edible films, kabog millet, fruit wastes

1. Introduction

1.1 Application of biodegradable films

In 2016, 335 million tons of plastics were produced globally, of which the majority were single-use plastics. Approximately 40% of the produced plastics were used for packaging [1]. This large volume of plastics leads to several problems: most oil-based plastics are highly resistant to biodegradation, which means if they enter the

IntechOpen

environment, they can accumulate, and this leads to adverse environmental consequences [1]. These negative side effects could be reduced with biodegradable plastics. Since biodegradable plastics can be produced from fruit and vegetable wastes, the use of biodegradable plastics also helps valorize food wastes. Creating biodegradable plastics from food wastes could reduce the carbon footprint and the adverse environmental effects of oil-based plastics. Packaging is an essential step in maintaining food quality during manufacturing and shelf life. Depending on the packaging properties, it can also protect from microorganisms and moisture. With proper packaging, shelf life can be extended, and food waste can be reduced [2]. Due to the crucial role of packaging in this context, it is essential to develop new biodegradable materials with desired properties. An interesting application would be a direct coating of a product with a thin layer of edible plastic. The coating could be enriched with antimicrobial substances or antioxidants, which protect the product directly [3].

1.2 Pectin from fruit wastes as an ingredient of edible films

Some biopolymers, such as casein, alginate, and pectin, are increasingly gaining attention because of their inherent biodegradability. Biopolymers made from polysaccharides have a higher thermostability compared to biopolymers based on proteins, such as casein [4]. To improve the properties of the bioplastic, combinations of different polysaccharides or additives can be used. An interesting polysaccharide that could be taken advantage of to form bioplastic is pectin. Pectin is one of the main structural polysaccharides in dicot plants and consists mainly of galacturonic acid units as sugar backbone [3]. The carboxyl groups of the uronic acid residues can be present in different degrees of methylesterification, influencing the processing properties [5]. Pectin is classified as high methoxyl pectin if more than 50% of the hydroxyl groups are esterified, and low methoxyl pectin if less than 50% of the hydroxyl groups are esterified [6]. A significant percentage of pectin is found in fruit peels that are often discarded. It is possible to use the discarded peels to create edible films, which could be applied as coating for fruits and vegetables. Additionally, unused fruit peel wastes can be reused by creating biodegradable plastic [4]. Pectin edible films might be used for the protection of food with low moisture content due to their low resistance to humidity. They also provide an excellent barrier to aroma compounds and oxygen [7]. Their poor water vapor barrier properties could be explained by the hydrophilic nature of pectin [7]. A plasticizer and polyvalent cations are often added to edible pectin film formulations [3]. Plasticizers are used to reduce the brittleness; thus, improving the mechanical properties of the film [8]. Polyvalent cations such as calcium can be applied to crosslink the pectin chains and form a gel [3].

1.3 Extraction methods for pectin from fruit substrates

There are multiple methods for extracting pectin from fruit wastes: for example, enzymatic, acidic, and microwave-assisted extraction. These methods differ in the yield of resulting pectin and its properties. One possibility is the application of an enzymatic preparation like Celluclast® 1.5 L. The enzymes partly degrade the plant cell wall components and enable the subsequent extraction of pectin [9]. The advantage of this method is that no strong acids are used; thus, no acidic wastes are produced. Microwave-assisted extraction is another way to extract pectin, characterized by its short processing time and improved yields. The yield of the extracted pectin varies with the microwave power used. The higher the applied microwave power, the higher the extraction yield [10].

1.4 Kabog millet flour as valuable ingredient of edible films

Kabog millet is an ecotype of *Panicum miliaceum* L. and grows only in the Philippines. The grain is rich in dietary fibers, proteins, and antioxidants, such as carotenoids and tocopherols [11]. Due to its high nutritional quality, the addition of whole-grain kabog millet flour to edible films is a promising option to improve their nutritional profile. In whole-grain kabog millet flour, lipids are also present, which may improve the hydrophobicity of the biopolymers [12]. Therefore, it may be a solution to one of the most common problems for bioplastics: low water resistance [13]. Whole-grain kabog millet flour has some promising properties that can be integrated into edible bioplastics formulations.

With these concepts in mind, the aims of this study were to 1) extract pectin from orange peel wastes by microwave-assisted extraction and enzymatic extraction; 2) characterize the extracted pectin by Fourier-transform infrared spectroscopy (FTIR) for the degree of esterification, by ion-exchange chromatography for the monosaccharide composition in pectin, and size-exclusion chromatography (SEC) for the polymer structure in water and its molecular weight; 3) create edible films containing kabog millet flour with the extracted and characterized pectin from different extraction methods; 4) characterize the produced films by their mechanical properties and their water contact angle; and 5) observe the effect of adding kabog millet flour on the properties of the films.

2. Materials and methods

2.1 Orange peel samples

The orange peel samples used to extract pectin were kindly provided by the Migros supermarket in Limmatplatz, Zürich (Switzerland) on February 07, 2022. The orange peels were produced as a by-product of orange juice production. They were frozen in plastic bags for 2 days at - 20°C and then freeze-dried for 2 days in a freeze dryer (Labconco FreeZone 4.5 Liter - 50°C Benchtop Freeze Dryer; Kansas City, US). Due to moisture after freeze-drying, the samples were cut into pieces with a diameter of 3 cm and were thoroughly dried in an oven at 70°C for 2 days. The dried orange peel pieces were shredded in a crusher (Durabase Migros; Zürich, Switzerland) and ground with a pestle and mortar. Prior to the pectin extractions, the moisture content was measured for 30 min at 120°C with a halogen moisture analyzer (Mettler Toledo HE53; Greifensee, Switzerland).

2.2 Chemicals

Hydrochloric acid (HCl) (≥37%), absolute ethanol (>99.8%), acetic acid (>99.8%), methanol (>99.9%), calcium chloride ($CaCl_2$) (≥ 97%, granulated), sodium acetate (NaOAc) (>99.0%), and Driselase (protein ≥10% from *Basidiomycetes* sp.) were purchased from Sigma-Aldrich GmbH (Steinheim, Germany). The Celluclast® 1.5 L enzymes were ordered from Novozymes (Bagsværd, Denmark). Glycerol (85%) was obtained from Hänsler AG (Herisau, Switzerland).

2.3 Microwave-assisted pectin extraction

The microwave-assisted extraction of pectin was performed according to Ref. [14] with a household microwave oven (Mio Star MW 01). Ten grams of sample powder was weighed in triplicates. Milli-Q water (pH 1.5) was added to the samples to reach an optimal liquid-to-solid ratio of 20 ml/g. The suspensions were microwaved at a power of 450 W and irradiation time of 20 min, and stirred every 5 min. The suspensions were then cooled down and transferred into 50-ml tubes. The tubes were centrifuged (Eppendorf 5810 R; Hamburg, Germany) at room temperature for 10 min at 4000 rpm and the supernatants were collected. The pectin was precipitated for 1.5 h at room temperature using an equal volume of 95% ethanol. The dispersion was filtered and washed three times with 95% ethanol at room temperature. The extracted pectin was dried at 50°C overnight.

2.4 Enzymatic extraction of pectin

The enzymatic extraction of pectin was performed according to [15], except that the duration was reduced from 18 h to 3 h. For extracting the pectin, 10 g of the orange peel powder was weighed in triplicates. To get a liquid-to-solid ratio of 15:1, 150 ml of Milli-Q water (pH 4.5) was used. For the suspension, 500 µl of Celluclast® 1.5 L enzymes was added (Novozymes, Bagsværd, Denmark) and the samples incubated for 3 h in a water bath at 50°C. The resulting solution was cooled to room temperature and transferred into 50-ml tubes. The tubes were centrifuged (Eppendorf 5810 R; Hamburg, Germany) at 4°C for 10 min at 4000 rpm. The supernatants were collected, and 96% ethanol was added at 4°C to a final ethanol concentration of 70%. The suspensions were precipitated for 1 h and transferred into 50-ml tubes, which were then centrifuged at 4000 rpm for 20 min at room temperature. The precipitate was washed with 70% ethanol and centrifuged again for 20 min under the same conditions. The combined precipitates were dried for 24 h at 60°C.

2.5 Characterization of the extracted pectins

2.5.1 Fourier-transform infrared spectroscopy (FTIR)

To determine the degree of esterification (DE) of the extracted pectin, FTIR analysis was performed according to Ref. [16], using an FTIR spectrometer (Varian 640 with golden gate-diamond ATR). To obtain a dense surface of pectin powder on the surface of the ATR diamond, three droplets of 100% methanol were added to the pectin powder to increase the density. After evaporation of the solvent, the absorption was measured several times to get the optimal spectrum with a definite pectin signal and a smaller signal of the remaining methanol. The DE could be calculated then with Eq. (1) using the absorbance (Abs) at the wavelengths at 1630 cm^{-1} and 1745 cm^{-1}, which are known as the fingerprint regions of pectin [16].

$$DE(\%) = \left[\mathrm{Abs}_{1745} / \left(\mathrm{Abs}_{1745} + \mathrm{Abs}_{1630} \right) \right] \times 100 \tag{1}$$

2.5.2 Ion-exchange chromatography

To analyze the pectin sample by ion-exchange chromatography, the pectin was digested with Driselase (Sigma-Aldrich) using a method of Ref. [17]. Five milligrams of the pectin samples was weighed in triplicates. For ion-exchange chromatography, a Thermo Scientific™ Dionex™ ICS-5000 (Thermo Fisher Scientific™, U.S.) was used with a PA1 Dionex CarboPac™ BioLc™ 4 x 50 mm column. The sampler had a 1 ml/min rate, and the reference electrode was AgCl. The eluting solvents were A: 200 mM NaOH, B: 100 mM NaOH +1 M NaOAc, and C: water. The gradient system was 7.6% A and 92.4% C for 0–26 min running time: 50% A and 50% C for 26–33 min: 47.2% A, 6% B, and 46.8% C for 33–45 min: 35.2% A, 30% B, and 34.8% C for 45–78 min: 100% B for 78–91 min: 50% A and 50% C for 91–99 min: and 7.6% A and 92.4% C for 99–105 min.

Two different kits from Megazyme were used to determine the free glucose and fructose in the pectin powder. The glucose kit K-GLUC and the glucose and fructose kit K-FRUGL were purchased from Megazyme Ltd. (Bray, Ireland), and the analyses were performed according to the manufacturer's protocols.

2.5.3 Size-exclusion chromatography (SEC)

For the sample preparation, 5 mg of the pectin samples was weighed in triplicates and dissolved in 5 ml of Milli-Q water. The solutions were kept for 1 h at 80°C while stirring in a water bath and afterwards at room temperature overnight while stirring. The transparent solutions were filtered through a nylon filter with pores of 0.45 µm (13 mm Syringe Filter, Nylon 66, 0.45 µm; BGB (Böckten, Switzerland)) into vials followed by analysis by size-exclusion chromatography (OMNISEC, Malvern Panalytical Ltd., Malvern, United Kingdom). The system consisted of an OMNISEC RESOLVE chromatography compartment combined with a pump, an autosampler and two A6000M columns in series (8.0 × 300 mm, OMNISEC REVEAL VISCOTEK, Malvern Panalytical Ltd., Malvern, United Kingdom). The OMNISEC RESOLVE detector compartment was equipped with a low and right-angle laser light scattering detector (LALS/RALS), a refractive index (RI), and a viscometer. The mobile phase was composed of a solution of 0.1 M $NaNO_3$ and contained 0.02% of NaN_3. Both columns were kept at 25°C, and the flow rate was 8.83 ml/min. The injection volume was 100 µl. A polyethyleneoxide (PEO-24 K, VISCOTEK, Malvern Panalytical Ltd., Malvern, United Kingdom) standard and a dextran standard (Dextran-T68K, American Polymer Standards Corporation, Mentor, US) were used for calibration. All vials were measured with three injections. The molecular weight (M_w), the intrinsic viscosity ([η]), the hydrodynamic radius (R_h), and the average conformation of the polymer (α) were determined by the instrument software.

2.6 Film preparation

The pectin films were prepared according to Ref. [18]. Six different prototypes of films were produced (**Table 1**). They differed in the extraction method of the pectin, pectin content, and the presence of whole-grain kabog millet flour (Catmon Cebu, Philippines). The whole-grain kabog millet flour was prepared according to Ref. [11].

The pectin was added carefully to the Milli-Q water with stirring. Glycerol, $CaCl_2$, and the whole-grain kabog millet flour (if present in the film formulation) were added. The suspensions were heated to 70°C, and 25 ml was poured into plastic weighing boats (Sigma-Aldrich GmbH, diameter = 13.5 cm). The films were dried overnight at room temperature and then in an oven at 29°C with 29% humidity until they were dry.

Code	Pectin (EN) [g]	Pectin (Mi) [g]	70% glycerol [g]	Milli-Q water [g]	$CaCl_2$ [mg]	Whole-grain kabog millet flour [g]
Mi1.5No	—	1.5	1.5	97.3	15	—
Mi1.5Millet	—	1.5	1.5	97.3	15	2
Mi2.5No	—	2.5	2.5	95	25	—
Mi2.5Millet	—	2.5	2.5	95	25	2
EN1.5No	1.5	—	1.5	97.3	15	—
EN1.5Millet	1.5	—	1.5	97.3	15	2

Table 1.
Ingredient composition of produced film prototypes (EN = enzymatic extraction, Mi = microwave-assisted extraction; No = without whole-grain kabog millet flour, millet = with whole-grain kabog millet flour).

2.7 Characterization of the pectin films

2.7.1 Mechanical tests

Mechanical properties of the films were evaluated by measuring tensile length and elongation using a Z010 (ZwickRoell GmbH & Co. KG, Ulm, Germany) with a 10 N load cell. The experiments were performed at room temperature and in triplicates using film pieces of 0.5 mm × 2.5 cm. The maximum strain and the stress to break were calculated according to Eqs. (2) and (3) using the elongation (L_F) from starting point of 1 cm, the breaking force (F) and the cross-sectional area (A).

$$\textit{Maximum strain}(\%) = \left[(L_F - 1\,\text{cm})/1\,\text{cm}\right] \times 100 \tag{2}$$

$$\textit{Maximum stress}(\text{MPa}) = F/A \tag{3}$$

2.7.2 Water contact angle measurement

To analyze the water contact angle of the different film prototypes, a Drop Shape Analyzer-DSA 100E (A.KRÜSS Optronic GmbH, Hamburg, Germany) was used. The contact angle of a 10 µl Milli-Q water droplet was determined by analyzing the drop shape. Measurements were done in triplicates at room temperature.

2.8 Statistical analysis

Data obtained from FTIR, the mechanical tests, and water contact angle measurements were evaluated with Microsoft Excel. For statistical analysis, an ANOVA was used followed by a post hoc Tukey's HSD ($p < 0.05$). The data obtained from ion-exchange chromatography, free sugar content analysis, and SEC were evaluated using IBM SPSS Statistics v. 28.0.0 (IBM, Armonk, United States) with a two-sided t-test, assuming homogenous variance ($p < 0.05$).

3. Results

3.1 Extraction yield

To compare the efficiency of the two different pectin extraction methods performed (enzymatic and microwave-assisted), the yields were calculated considering the moisture content (MC) of the orange peel powder used ($MC_{Mi} = 5.85\%$, $MC_{EN} = 6.36\%$). The yield of the microwave-assisted extraction (Mi) was 15.07 ± 0.91% and differed significantly from the enzymatic extraction yield (EN) (3.11 ± 0.40%, $n = 3$ and $p < 0.05$).

3.2 Degree of esterification (DE): FTIR

To characterize the different extracted pectin samples, three of them were tested on their degree of esterification: the two sample sets used for measuring extraction yield of the enzymatic and the microwave-assisted extraction methods, and pectin extracted by microwave in bulk amounts, used later for film production. EN pectin had a DE = 49.10 ± 2.66% and was significantly different from the Mi pectin (DE = 77.34 ± 0.88%) and exhibited a significant difference from the bulk Mi pectin (**Table 2**).

The Mi pectin was considered highly esterified in both cases since the degree of esterification was greater than 50%. The EN pectin could be described as moderately esterified because the resulted degree of esterification was located at the boundary of high- and low-esterified pectins.

Sample name	DE [%]
Mi pectin	$77.34^{b} \pm 0.88$
EN pectin	$49.10^{c} \pm 2.66$
Bulk Mi pectin	$85.21^{a} \pm 3.20$

Data are presented as mean ± standard deviation (n = 3). Different subscript letters indicate significantly different values at p < 0.05 using Tukey's HSD.

Table 2.
DE of pectin extracted by different methods.

	Amount in Mi pectin [%]	Amount in EN pectin [%]	
L-Fucose	4.14 ± 0.44	3.35 ± 0.19	
L-Rhamnose	5.56 ± 0.45	3.60 ± 0.28	*
L-Arabinose	11.47 ± 0.23	7.32 ± 0.28	*
D-Galactose	9.86 ± 0.35	12.41 ± 0.40	*
D-Xylose	4.29 ± 0.44	5.37 ± 0.78	
D-Galacturonic acid	64.68 ± 1.65	67.94 ± 1.74	

Data are presented as mean ± standard deviation (n = 3). Asterisks show significant differences between the two extraction methods at p < 0.05 using a two-sided t-test with the same variance.

Table 3.
Monosaccharide composition of pectin differing in the extraction method without glucose.

3.3 Pectin composition: ion-exchange chromatography

To compare the monosaccharide compositions of the differently extracted pectin samples, ion-exchange chromatography was performed. Prior to this analysis, the pectin samples were digested with Driselase. The monosaccharide compositions of both pectin samples were comparable. For both extraction methods, the main monosaccharide was galacturonic acid, which did not differ significantly between the two extraction methods. Rhamnose, arabinose, and galactose differed significantly (**Table 3**).

The undigested pectin powders were also tested for their free glucose and fructose contents, using two different Megazyme kits. The K-GLUC kit measured only the free glucose and the K-FRUGL measured the free glucose as well as the free fructose. It showed that the pectin powder extracted from orange peel contained some free sugar such as fructose and glucose. The Mi pectin contained significantly more free sugars than EN pectin (**Table 4**).

3.4 Size-exclusion chromatography

The different pectin samples were analyzed by SEC to test the behavior of the molecules in aqueous solutions and to get information about their particle size. The molecular weight (M_w), the intrinsic viscosity ($[\eta]$), the hydrodynamic radius (R_h), and the average conformation of the polymer (α) were determined. The chromatograms of the two pectin samples each showed two populations of molecules, which were quantified. The two peaks were overlapping in the chromatogram, leading to approximate integration.

The two populations of molecules from the two pectin samples analyzed differed significantly in their molecular weight. The molecules extracted using the microwave-assisted procedure were characterized by higher molecular weight than those of the other pectin sample (**Table 5**). The chromatogram showed that Mi pectin had higher dispersity in molecules than EN pectin. The intrinsic viscosity of the different peaks varied significantly. The first peak of the EN pectin had the highest intrinsic viscosity (8.73 ± 0.39 dl/g). The intrinsic viscosity was higher at the first peaks, implying a more open structure and a higher hydrodynamic radius. The average conformation of both molecule populations of the EN pectin was significantly higher than that of the other pectin samples (**Table 5**).

3.5 Mechanical tests of the pectin films

The mechanical properties of the six pectin-containing film prototypes (for preparation see **Table 1**, Section **2.6**) were determined in triplicates. The parameters from these

	Free glucose [mg/l] (K-GLUC)	**Free glucose [mg/l] (K-FRUGL)**	**Free fructose [mg/l] (K-FRUGL)**
EN pectin	33.26 ± 3.70*	20.96 ± 2.03*	11.44 ± 3.77*
Mi pectin	92.85 ± 2.20*	34.77 ± 2.93*	28.73 ± 3.32*

Data are presented as mean ± standard deviation (n = 3). Asterisks show significant differences between the two extraction methods at p < 0.05 using a two-sided t-test with the same variance.

Table 4.
Free glucose and fructose measurement by two different Megazyme kits and UV/Vis spectroscopy.

	EN pectin	Mi pectin	
M_w Peak 1 [kDa]	211.10 ± 18.09	484.61 ± 15.84	*
M_w Peak 2 [kDa]	67.60 ± 2.87	90.55 ± 5.59	*
[η] Peak 1 [dl/g]	8.73 ± 0.39	6.46 ± 0.25	*
[η] Peak 2 [dl/g]	0.66 ± 0.10	0.98 ± 0.16	*
R_h Peak 1 [nm]	29.91 ± 1.31	35.35 ± 0.86	*
R_h Peak 2 [nm]	8.39 ± 0.47	11.07 ± 0.77	*
α Peak 1	0.983 ± 0.034	0.596 ± 0.020	*
α Peak 2	1.006 ± 0.227	N/C	

Data are presented as mean ± standard deviation (n = 9). Asterisks show significant differences between the two extraction methods at p < 0.05 using a two-sided t-test with same variance.

Table 5.
Molecular weight, intrinsic viscosity, hydrodynamic radius, and average conformation of the two pectin samples analyzed by OMNISEC with PEO-24 K as a calibration standard.

measurements, maximum strain and maximum stress, are shown in **Figures 1** and **2**, respectively. The maximum strain of the film Mi2.5No was 138.26 ± 13.97% and differed significantly from Mi1.5Millet, Mi2.5Millet, and EN1.5No, with considerably lower maximum strains than Mi2.5No. The films EN1.5Millet and Mi1.5No showed intermediate maximum strains and were not significantly different from the data of all other prototypes (**Figure 1**).

At the maximum stress, a significant difference between films produced with and without the whole-grain kabog millet flour was visible (**Figure 2**). The maximum stress of the film Mi2.5Millet was 3.50 ± 0.33 MPa, and the one from the film Mi2.5No was 8.38 ± 0.61 MPa. The only difference in the composition of these two films was the addition of 2 g whole-grain kabog millet flour to Mi2.5Millet, resulting in more brittle film properties. This observation was also significant between EN1.5No and EN1.5Millet and between Mi1.5No and Mi1.5Millet (**Figure 2**).

To determine the hydrophobicity of the created films, the static water contact angle was measured on the surface of the films. The films showed significant differences in

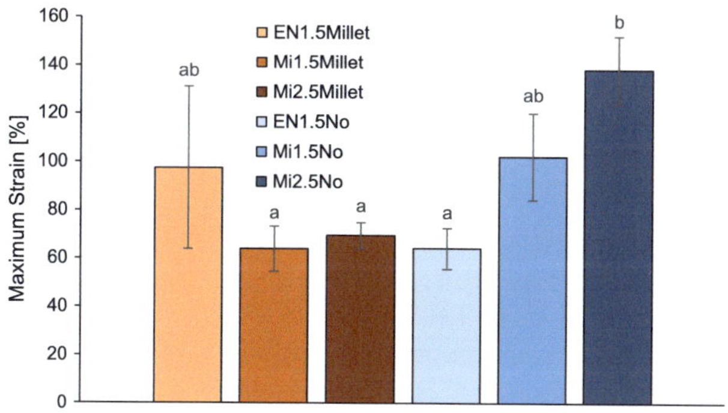

Figure 1.
The maximum strain of selected films. The different codes stand for different film prototypes compositions: EN = enzymatic extraction of pectin; Mi = microwave-assisted extraction of pectin; 1.5 = contains 1.5 g extracted pectin; 2.5 = contains 2.5 g extracted pectin; No = without whole-grain kabog millet flour; Millet = with whole-grain kabog millet flour. Data are presented as mean ± standard deviation (n = 3). Different subscript letters indicate significantly different values at p < 0.05 using Tukey's HSD.

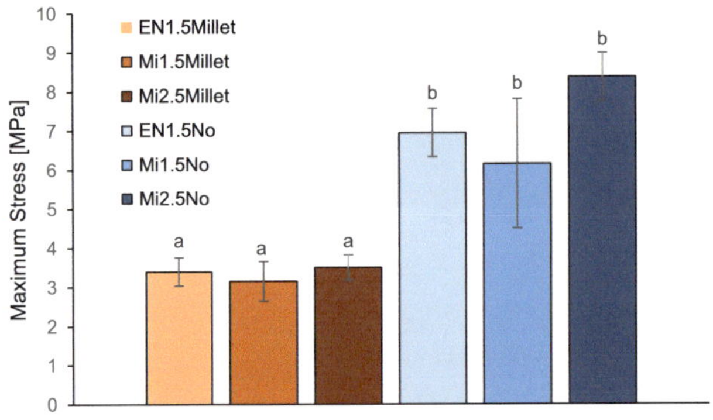

Figure 2.
The maximum stress of selected films. The different codes stand for different film prototypes compositions: EN = enzymatic extraction of pectin; Mi = microwave-assisted extraction of pectin; 1.5 = contains 1.5 g extracted pectin; 2.5 = contains 2.5 g extracted pectin; No = without whole-grain kabog millet flour; and Millet = with whole-grain kabog millet flour. Data are presented as mean ± standard deviation (n = 3). Different subscript letters indicate significantly different values at p < 0.05 using Tukey's HSD.

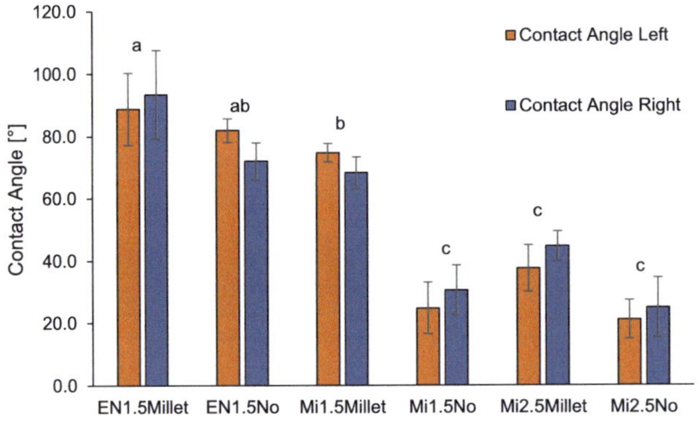

Figure 3.
Water contact angle left and right of selected films. The different codes stand for different film prototypes compositions: EN = enzymatic extraction of pectin; Mi = microwave-assisted extraction of pectin; 1.5 = contains 1.5 g extracted pectin; 2.5 = contains 2.5 g extracted pectin; No = without whole-grain kabog millet flour; and Millet = with whole-grain kabog millet flour. Data are presented as mean ± standard deviation (n = 3). Different subscript letters indicate significantly different values at p < 0.05 using Tukey's HSD.

contact angle as a function of the extraction method of pectin, the pectin concentration, and the presence of whole-grain kabog millet flour (**Figure 3**). Based solely on the extraction method of the pectin, it was observed that films produced with EN pectin showed significantly higher water contact angle than films produced from Mi pectin. This could be observed by comparing the mean contact angle of EN1.5No (76.94° ± 4.90) and Mi1.5No (27.68° ± 8.15). Biopolymers produced from EN pectin were characterized by higher hydrophobicity compared to biopolymers made from Mi pectin.

Comparing the difference in water contact angle between Mi2.5No/Millet and Mi1.5No/Millet, it could be observed that the change in the contact angle by adding the whole-grain kabog millet flour was decreased. With increasing pectin content, the effect of the whole-grain kabog millet flour on hydrophobicity was significantly reduced, and overall hydrophilicity was enhanced significantly. Comparing the water contact angle of Mi1.5Millet at 71.48° ± 4.05 and of Mi1.5No at 27.68 ± 8.15,

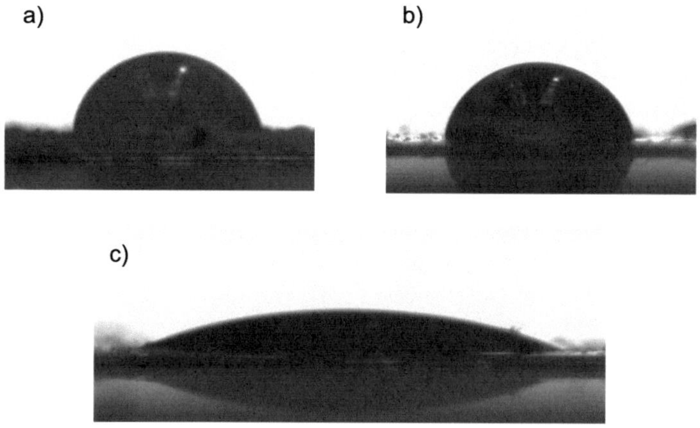

Figure 4.
Water droplet on the surface of a) Mi1.5Millet b) EN1.5No c) Mi1.5No. The different codes stand for different film prototypes compositions: EN = enzymatic extraction of pectin; Mi = microwave-assisted extraction of pectin; 1.5 = contains 1.5 g extracted pectin; No = without whole-grain kabog millet flour; Millet = with whole-grain kabog millet flour.

a significant increase in contact angle, and therefore in hydrophobicity could be observed with adding whole-grain kabog millet flour to the biopolymer. In comparison with the other two pairs of films, there was no significant effect observed from the addition of whole-grain kabog millet flour to the film. It is visible that the water contact angle was significantly lower for Mi1.5No compared to the films with enzymatically extracted pectin (EN1.5No) or added kabog millet flour (Mi1.5Millet) (**Figure 4a–c**).

4. Discussion

4.1 Extraction yield

The extraction yield differed significantly between the two extraction methods. The yield from enzymatic extraction using Celluclast® 1.5 L (~3%) was significantly lower than other studies using a similar protocol for apple pomace [19]. Orange peels contain 20.9% pectin [20] and apple pomace, 19–20% [21]. Therefore, the source of the pectin was not the main reason for the difference in yield. A possible explanation for the low yield in the current study is the incubation time of 3 h, which was shorter than 18 h used by Ref. [19]. It can be assumed that more cell wall components are digested with longer incubation time, and more pectin can be extracted [19]. Yield could be increased in further experiments by using longer incubation time or increasing the enzyme concentration. The yield from microwave-assisted extraction (~15%) was comparable to other studies [22]. The presence of free sugar, especially in the Mi pectin powder, can contribute to errors in the gravimetric measurement of the yield. A possible source of error could be also the precipitation of other polymers, which can bind to pectin, such as cellulose. Cellulose can also be extracted from plant material using microwave-assisted extraction and ethanol as solvent [23]. Therefore, it is possible that other polymers present in plant cell walls can precipitate and affect the gravimetric measurement of the pectin extraction yield.

4.2 Degree of esterification

The DE differed significantly between the two extraction methods. EN pectin can be described as moderately esterified because its DE was just at the boundary between high- and low-esterified. The obtained DE (~49%) was comparable to DE values using the same enzyme for extracting pectin [15]. It may be possible that only pectin with a lower degree of esterification can be released from the plant cell by enzymatic treatment. The DE of the microwave-assisted extracted pectin could be considered as highly esterified (~77%). The result was comparable to that of microwave-assisted extracted orange peel pectin (~71%) from [24], from lime peels (~71–92%) [25], and apple pomace pectin extracted by microwave (~74%) [9].

4.3 Pectin composition

As expected, the monosaccharide compositions of both pectin samples were comparable because the pectin was extracted from the same source and should therefore have a similar composition. The main component was galacturonic acid, which is the backbone molecule of pectin [4]. The galacturonic acid contents from both Mi pectin and EN pectin did not differ significantly. It was comparable to the galacturonic acid content in apple pectin [19] and orange peel pectin [24, 26]. The other monosaccharides except for rhamnose, galactose, and arabinose were also similar. The three significantly differing monosaccharides could be converted during extraction. The free glucose test with the K-GLUT assay was significantly higher in the Mi pectin powder than in the enzymatically extracted one. Due to the differences in the free glucose testing between the two applied kits, it can be assumed that the results had some uncertainties. These could be explained by the approximation of the reaction endpoint used in the K-FRUGL kit. Another reasonable explanation would be the presence of cellulose in the extracted pectin powder. (Nano) cellulose fibrils can be extracted using microwave-assisted methods [23]. Considering that orange peel has a cellulose content of approximately 50%, it is reasonable that the pectin powder also contained nanocellulose molecules [27]. Since microwave-assisted extraction is more vigorous and can increase the solubility of certain cellular compounds, it can be explained that more free glucose and nanocellulose fibrils were extracted from the cell wall and the glucose content, therefore, was higher in the Mi pectin sample than in the EN pectin powder.

4.4 Pectin molecular properties

The observed molecular weights of the first peak of each pectin sample from enzymatic extraction and microwave-assisted extraction were ~ 211 ± 18 kDa and ~ 485 ± 20 kDa, respectively. These values were comparable to the molecular weights of orange peel pectins, which had a molecular weight of 120 ± 10 kDa to 360 ± 20 kDa [28]. The dispersion of particle size and, therefore, of the molecular weight was broader in microwave-assisted extraction due to the rapid increase in temperature and internal pressure [16]. The second peaks could depict cleaved pieces of pectin or other dissolved cell components like cellulose. There are cellulose molecules or cellulose derivatives that would fit in the range of the molecular weight of the second peaks [29]. The intrinsic viscosity values from enzymatic extraction (~8.7 dl/g) and microwave-assisted extraction (~6.5 dl/g) fit within the range of 4.8–10.8 dl/g for the intrinsic viscosity of orange peel pectin [28]. The Mi pectin was characterized by a significantly lower intrinsic viscosity than the EN pectin. A lower intrinsic viscosity describes a more condensed structure and higher molecular density [30]. In contrast to the intrinsic viscosity, the Mi pectin had a significantly higher hydrodynamic radius than the EN pectin, indicating a looser conformation. Due to this contradictory relationship, it can be assumed that the condensation of the pectin structures in water is similar. The average conformation of the two pectin samples differed. The EN pectin had values nearly twice as high as the Mi pectin. The Mi pectin had a conformation that can be described as a semi-flexible random coil-like structure due to its α value between 0.5 and 0.8 [30]. On the other hand, the EN pectin had both populations of particles with an average conformation higher than 0.8. They tend to be in a more rigid and rod-like structure.

4.5 Mechanical testing of the created films

The differences in maximum strain showed the trend that the elongation could be enhanced by increasing the pectin content. The maximum strain of Mi1.5No was 102.25 ± 17.93%, and if increasing the pectin content by 1 g/100 ml, the maximum strain reached 138.26 ± 13. 97% (Mi2.5No). The maximum stress was therefore increased by approximately 36%, which can be explained by increased film thickness and consequently, increased cross-sectional area; thus, more cross-linkages that need to be broken [18]. The film microstructure was altered with changing pectin content by analyzing with scanning electron microscopy [18]. With 2.5 g pectin per 100 ml film-forming solution, the film structure was the smoothest [18]. By increasing or decreasing the pectin content, the microstructure was disturbed. A significant effect of the different extraction procedures on the maximum strain of the film could not be observed. A significant change in maximum strain by adding whole-grain kabog millet flour could only be observed between the prototypes Mi2.5No (138.26 ± 13.97%) and Mi2.5Millet (69.45 ± 5.50%). The higher maximum strain of Mi2.5No cannot be explained by increasing the cross-sectional area because the films containing whole-grain kabog millet flour were characterized by the increased film thickness. It could be assumed that the internal arrangement was disturbed by adding the kabog millet flour; therefore, the elongation properties decreased. Compared to other biodegradable films based on polymers, the produced films tended to be rigid and inelastic. There were formulations for biodegradable films characterized by a maximum strain between 300% and 700% [13]. Regarding the maximum mechanical stress to break, films with kabog millet flour broke with the application of approximately half of the force per cross-sectional area than films without kabog millet flour. This leads to the conclusion that adding kabog millet flour changed the properties of the films toward higher brittleness. No significant differences between the maximum stress data of the films produced from different pectin samples could be observed. Therefore, the extraction method did not influence the mechanical properties of the produced pectin edible films.

4.6 Water contact angle

The water contact angles on the surface of the produced films provided information about the hydrophobic characteristics of the different film prototypes. A low water contact angle characterizes more hydrophilic properties of the edible pectin film [13]. In the application of pectin films for covering food products, a certain hydrophobicity would be desired to stabilize the protection layer against water. It could be observed that the films produced from EN pectin resulted in higher water contact angles than films from the Mi pectin sample. This is contradictory to the results of the degree of esterification because the EN pectin showed a greater number of free carboxy groups due to its lower degree of esterification and, therefore, should be more hydrophilic [31]. A possible explanation for the lower contact angles observed for the Mi pectin could be its higher hydrodynamic radius. The higher hydrodynamic radius indicates higher hydrophilicity. Furthermore, the rod-like conformation of the EN pectin does not allow as many interactions with the surrounding water because only a part of the functional groups is exposed on the surface of the rod and therefore is characterized as more hydrophobic. The addition of the whole-grain kabog millet flour showed a significant increase in water contact angle between the films Mi1.5Millet and Mi1.5No, while between the other pairings, a difference could be observed but it was less pronounced. The addition of the kabog millet flour increased the hydrophobicity, which could be explained by its whole-grain character. In whole-grain products, lipids are present, including some lipophilic compounds such as carotenoids and tocopherols [11]. These lipophilic structures may shift and enhance the hydrophobic properties of the films. During the measurements of the water contact angle, some properties of the film were observed, which could be interesting for further research. The pectin films, especially the ones with kabog millet, showed volume expansion upon the addition of water, and the droplet of water turned white over time, so it can be assumed that some hydrophilic film compounds migrated into the water droplet.

5. Conclusion and outlook

The properties of the pectin obtained from enzymatic and microwave-assisted extraction differed in several of the tested properties. Enzymatically-extracted pectin was characterized as moderately esterified (49.1% DE) with a rod-like conformation in water and with a lower average molecular weight than the microwave-assisted extracted pectin. These pectin samples had a random coil-like conformation and were highly esterified (>75% DE). In addition, the yields of the two extraction methods were significantly different. While the microwave-assisted extraction had a yield comparable to literature, the enzymatic extraction resulted in low yield which may be improved in further experiments by increasing the enzyme concentration, incubation time, or other relevant parameters. The properties of films prepared from the extracted pectin material showed significant differences. Overall, the elasticity of the films was considerably lower than the data of comparable material described in the literature. The high amount of presumably cellulose (>70% glucose monomer units) present in the extracted material might play a role for the low elasticity observed. The enzymatically-extracted pectin is a promising material to produce edible pectin films due to the high contact angles determined for the films and thus enhanced hydrophobicity and stability to water. Adding whole-grain kabog millet flour increased the hydrophobic properties of the films but concomitantly, the brittleness of the pectin edible films also increased. It would need further research to adjust the protocol to produce pectin films containing whole-grain kabog millet flour without changing their elasticity. Furthermore, it would be interesting to evaluate the antioxidant properties of compounds present in the whole-grain kabog millet applied in a film or coating. This could pave the way for fruit waste valorization and the addition of nutritional components to edible, biodegradable films.

Acknowledgements

We are grateful for the invaluable support of Serena Rosa Alfarano for the FTIR measurements, Horst Adelmann for the halogen moisture analyzer and the mechanical tests, Mohammad Peydayesh for sharing his experience in producing and testing plastic films, Tonghui Jin for the water contact angle measurements, and Cristina Lupo for the SEC and data analyses.

Author contributions

Nils Rentsch: Conceptualization, data curation, formal analysis, investigation, validation, visualization, writing—original draft; Laura Nyström: Project administration, resources, software, writing—review and editing; Joan Oñate Narciso: Conceptualization, data curation, formal analysis, funding acquisition, investigation, methodology, supervision, validation, writing—review and editing.

Funding

This research was generously and kindly supported by the Future Food Initiative (Future Food Fellowship), ETH Zürich World Food System Center, ETH Zürich Foundation project number: ETH 2019-HE-322. The funding source was not involved in the study design, in the collection, analysis and interpretation of data, in the writing of the report, and in the decision to submit the manuscript for publication.

Declaration of competing interest

None.

Author details

Nils Rentsch, Laura Nyström and Joan Oñate Narciso*
Laboratory of Food Biochemistry, Department of Health Sciences and Technology, ETH Zürich, Zürich, Switzerland

*Address all correspondence to: joan.narciso@hest.ethz.ch

References

[1] Narancic T, O'Connor KE. Plastic waste as a global challenge: Are biodegradable plastics the answer to the plastic waste problem? Microbiology. 2019;**165**(2):129-137. DOI: 10.1099/mic.0.000749

[2] Ashfaq J, Channa IA, Shaikh AA, Chandio AD, Shah AA, Bughio B, et al. Gelatin-and papaya-based biodegradable and edible packaging films to counter plastic waste generation. Materials. 2022;**15**(3):1046. DOI: 10.3390/ma15031046

[3] Mohamed SA, El-Sakhawy M, Abdel-Monem E-SM. Polysaccharides, protein and lipid-based natural edible films in food packaging: A review. Carbohydrate Polymers. 2020;**238**:116178. DOI: 10.1016/j.carbpol.2020.116178

[4] Mellinas C, Ramos M, Jiménez A, Garrigós MC. Recent trends in the use of pectin from Agro-waste residues as a natural-based biopolymer for food packaging applications. Materials. 2020;**13**(3):673. DOI: 10.3390/ma13030673

[5] Monsoor MA, Kalapathy U, Proctor A. Improved method for determination of pectin degree of esterification by diffuse reflectance Fourier transform infrared spectroscopy. Journal of Agricultural and Food Chemistry. 2001;**49**(6):2756-2760

[6] Iijima M, Nakamura K, Hatakeyama T, Hatakeyama H. Phase transition of pectin with sorbed water. Carbohydrate Polymers. 2000;**41**(1):101-106. DOI: 10.1016/S0144-8617(99)00116-2

[7] Giancone T, Torrieri E, di Pierro P, Cavella S, Giosafatto CVL, Masi P. Effect of surface density on the engineering properties of high Methoxyl pectin-based edible films. Food and Bioprocess Technology. 2011;**4**(7):1228-1236. DOI: 10.1007/s11947-009-0208-9

[8] Fishman ML, Coffin DR. Mechanical, microstructural and solubility properties of pectin/poly(vinyl alcohol) blends. Carbohydrate Polymers. 1998;**35**(3-4):195-203. DOI: 10.1016/S0144-8617(97)00245-2

[9] Dranca F, Oroian M. Optimization of pectin enzymatic extraction from *Malus domestica* "Fălticeni" apple pomace with Celluclast 1.5L. Molecules. 2019;**24**(11):2158. DOI: 10.3390/molecules24112158

[10] Yu XC, Sun DL. Microwave and enzymatic extraction of orange peel pectin. Asian Journal of Chemistry. 2013;**25**(10):5333-5336. DOI: 10.14233/ajchem.2013.14175

[11] Narciso JO, Nyström L. Breathing new life to ancient crops: Promoting the ancient Philippine grain "kabog millet" as an alternative to rice. Food. 2020;**9**(12):1727. DOI: 10.3390/foods9121727

[12] Hao B, Narciso JO, Nyström L. Extraction and identification of phospholipids from whole grain kabog millet flour and predictive effects on starch binding and retrogradation. LWT. 2021;**152**:112406. DOI: 10.1016/j.lwt.2021.112406

[13] Peydayesh M, Bagnani M, Mezzenga R. Sustainable bioplastics from amyloid fibril-biodegradable polymer blends. ACS Sustainable Chemistry & Engineering. 2021;**9**(35):11916-11926. DOI: 10.48550/arXiv.2105.14287

[14] Wang S, Chen F, Wu J, Wang Z, Liao X, Hu X. Optimization of pectin

extraction assisted by microwave from apple pomace using response surface methodology. Journal of Food Engineering. 2007;**78**(2):693-700. DOI: 10.1016/j.jfoodeng.2005.11.008

[15] Wikiera A, Mika M, Starzyńska-Janiszewska A, Stodolak B. Application of Celluclast 1.5L in apple pectin extraction. Carbohydrate Polymers. 2015;**134**:251-257. DOI: 10.1016/j.carbpol.2015.07.051

[16] Karbuz P, Tugrul N. Microwave and ultrasound assisted extraction of pectin from various fruits peel. Journal of Food Science and Technology. 2020;**58**:641-650. DOI: 10.1007/s13197-020-04578-0

[17] Zhang T, Tang H, Vavylonis D, Cosgrove DJ. Disentangling loosening from softening: Insights into primary cell wall structure. Plant Journal. 2019;**100**(6):1101-1117. DOI: 10.1111/tpj.14519

[18] Galus S, Turska A, Lenart A. Sorption and wetting properties of pectin edible films. Czech Journal of Food Sciences. 2012;**30**(5):446-455. DOI: 10.17221/444/2011-CJFS

[19] Wikiera A, Mika M, Grabacka M. Multicatalytic enzyme preparations as effective alternative to acid in pectin extraction. Food Hydrocolloids. 2015;**44**:156-161. DOI: 10.1016/j.foodhyd.2014.09.018

[20] Rivas-Cantu RC, Jones KD, Mills PL. A citrus waste-based biorefinery as a source of renewable energy: Technical advances and analysis of engineering challenges. Waste Management and Research. 2013;**31**(4):413-420. DOI: 10.1177/0734242X13479432

[21] Cinkmanis I, Muizniece-Brasava S, Viluma I, Vucane S, Aboltins A, Keke A. Extraction of pectin from apple pomace. Engineering for Rural Development. 2020;**19**:1934-1939. DOI: 10.22616/erdev.2020.19.tf549

[22] Prakash Maran J, Sivakumar V, Thirugnanasambandham K, Sridhar R. Optimization of microwave assisted extraction of pectin from orange peel. Carbohydrate Polymers. 2013;**97**(2):703-709. DOI: 10.1016/j.carbpol.2013.05.052

[23] Asada C, Katsura K, Suzuki A, Nakamura Y. Extraction, separation, and utilization of components contained in waste bamboo by pressurized microwave-assisted ethanol solvent treatment. Biomass Conversion and Biorefinery. 2021;**1**:3. DOI: 10.1007/s13399-021-01781-7

[24] Su DL, Li PJ, Quek SY, Huang ZQ, Yuan YJ, Li GY, et al. Efficient extraction and characterization of pectin from orange peel by a combined surfactant and microwave assisted process. Food Chemistry. 2019;**286**:1-7. DOI: 10.1016/j.foodchem.2019.01.200

[25] Rodsamran P, Sothornvit R. Microwave heating extraction of pectin from lime peel: Characterization and properties compared with the conventional heating method. Food Chemistry. 2019;**278**:364-372. DOI: 10.1016/j.foodchem.2018.11.067

[26] Kar F, Arslan N. Characterization of orange peel pectin and effect of sugars, L-ascorbic acid, ammonium persulfate, salts on viscosity of orange peel pectin solutions. Carbohydrate Polymers. 1999;**40**(4):285-291. DOI: 10.1016/S0144-8617(99)00063-6

[27] Bicu I, Mustata F. Cellulose extraction from orange peel using sulfite digestion reagents. Bioresource Technology. 2011;**102**(21):10013-10019. DOI: 10.1016/j.biortech.2011.08.041

[28] Fishman ML, Chau HK, Hoagland P, Ayyad K. Characterization of pectin, flash-extracted from orange albedo by microwave heating, under pressure. Carbohydrate Research. 1999;**323**(1-4):126-138. DOI: 10.1016/S0008-6215(99)00244-X

[29] Goodwin DJ, Picout DR, Ross-Murphy SB, Holland SJ, Martini LG, Lawrence MJ. Ultrasonic degradation for molecular weight reduction of pharmaceutical cellulose ethers. Carbohydrate Polymers. 2011;**83**(2):843-851. DOI: 10.1016/j.carbpol.2010.08.068

[30] Lupo C, Boulos S, Nyström L. Influence of partial acid hydrolysis on size, dispersity, monosaccharide composition, and conformation of linearly-branched water-soluble polysaccharides. Molecules. 2020;**25**(13):2982. DOI: 10.3390/molecules25132982

[31] Wang P, Fei P, Zhou C, Hong P. Stearic acid esterified pectin: Preparation, characterization, and application in edible hydrophobic pectin/chitosan composite films. International Journal of Biological Macromolecules. 2021;**186**:528-534. DOI: 10.1016/j.ijbiomac.2021.06.030

Chapter 2

New Sources of Pectin: Extraction, Processing, and Industrial Applications

Stefani Cortés-Camargo, Angélica Román-Guerrero, Erik Alpizar-Reyes and César Pérez-Alonso

Abstract

One of the most important polysaccharides in the vegetal kingdom is pectin. This class of natural polysaccharide is found primarily in citrus fruits and apple pomace. Pectin has been used in different sectors of the industry, among which the food, pharmaceutical, cosmetic, and paper industries stand out. Today, there is a growing demand for this type of hydrocolloid, where both the scientific and industrial fields have focused on using new sources of pectin and developing novel extraction methods. This chapter describes the chemical structure of pectin and its main chemical characteristics. Then, the conventional sources from which pectin is obtained are exposed as well as its main industrial applications. Subsequently, the physicochemical and functional properties of pectins obtained from unconventional sources are described and analyzed as well as the main technologies used for their extraction. Finally, the most recent advances in the role played by pectin in the industrial sector are described.

Keywords: pectin, extraction, functional properties, husks, hulls, Cactaceae, new applications

1. Introduction

Pectin is considered one of the main polysaccharides found in plant sources; it participates in the constitution of cell walls of higher plants, impacting the physical and nutritional contribution of products of plant origin. Pectin is a globally recognized polysaccharide with great relevance in the global biopolymer market due to its inherent functional properties and vast applications in the food, pharmaceutical, and biomedical industries [1]. It is a macromolecule capable of forming flexible polymer chains that lead to forming hydrogel-type structures [2]. Its functional properties are associated with the extraction conditions and influenced by the source used. The primary sources of commercial pectin are citric fruits and apples; however, non-conventional sources have been investigated, such as agro-industrial sub-products and residues, pulps, husks, hulls, peels, Cactaceae, and vegetables, among others [3]. Furthermore, pectin has been functionalized through chemical or enzyme reactions that lead to changes and improvements in its physicochemical properties, such as molecular weight, degree of esterification (DE), and surface charge, which in turn contributes to the development of new functional or improved properties, along with new approaches and applications [4].

2. Pectin structure

Pectin is a negatively charged branched heteropolysaccharide, composed of up to 17 different monosaccharides with more than 20 types of linkages [5, 6]. This polysaccharide was first reported in 1825 by Braconnot and defined as a biopolymer rich in galacturonic acid (GalA; up to 65%) [7]. Although the precise structure of pectin has not yet been fully elucidated due to its complexity, three major polysaccharide domains are recognized; as shown in **Figure 1**, the most abundant is based on a linear homopolymer of α-(1–4)-linked-D-galacturonic acid (GalpA, GalA) residues that can be methyl esterified at the C-6 position and to a lesser extent O-acetylated in C-2 and C-3; this domain is defined as homogalacturonan (HG) [5, 7]. In the rhamnogalacturonan I (RG-I) domain, the rhamnose (Rhap, Rha) residues disrupt the HG structure to form a preferably ramified structure of pectin (20–35%) due to the presence of the repeating disaccharide [→4)-α-D-GalpA-(1→2)- α-L-Rhap-(1-]. Here, the GalA residues are not methyl esterified, and attachment of neutral sugar side chains [α-L-arabinose (Araf, Ara) and β-D-galactose (Galp, Gal)] to the C-4 positions of Rha residues can be suitable, leading to linear side chains (LC-A) when α(1→5)-L-Araf or linear type I (β(1→4)-L- Galp) or branched side chains (LC-B) when α(1→2,3)-L-Araf or branched

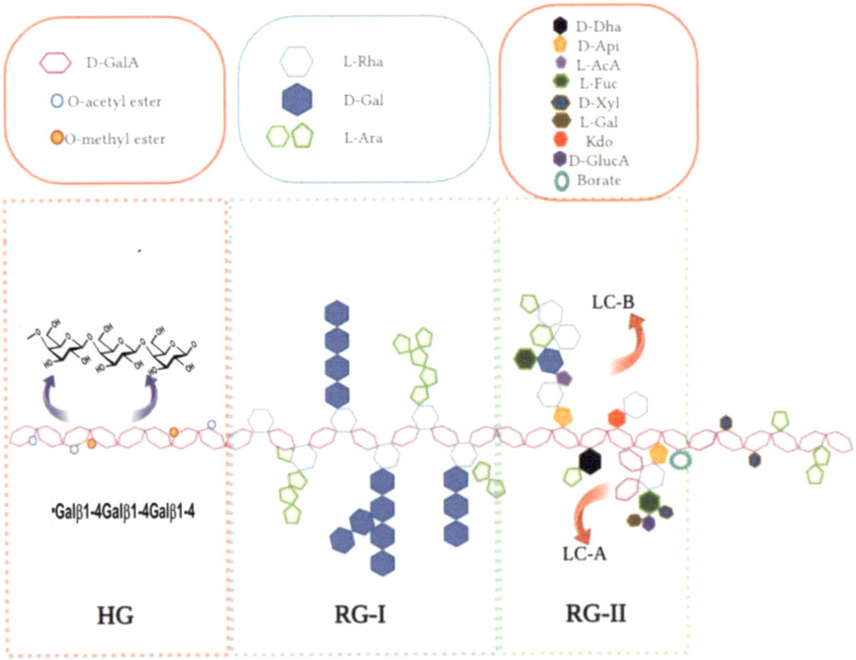

Figure 1.
The schematic representation of the pectin structure contains the HG, RG-I, and RG-II domains. L-AcA: L-Aceric acid. Adapted from [8].

type II β(1→3,6)-D-Galp and arabinogalactans. The branching design of the structure in RG-I depends on the pectin source, the extraction conditions, and the presence of other sugars such as xylose (D-Xyl), fucose (L-Fuc), and glucuronic acid (D-GlucA), among others [9]. The RG-II domain (1–8%) is constituted of around nine α(1→4)-linked GalpA units partially methyl esterified with four heteropolymer side chains attached, mainly composed of 11 monosaccharide residues, including apiose (D-Api), 2-O-methyl-L-fucose, 2-O-methyl-D-xylose. 3-C-carxy-5deoxy-L-xylose, 3-deoxy-D-manno-octulosonic acid (Kdo), and 3-deoxy-D-lyxoheptulosaric acid (D-Dha), which are linked with up to 22 glycoside bonds [10, 11].

Some investigations about the basic structure of pectin establish that although the pectin source may influence the structure diversity by partially modifying the chain conformation of the macromolecule, the RG-II region seemed to be well preserved among the different sources [12]. Moreover, pectins contain functional groups besides carbohydrate type, such as phenolic acids, methanol, acetic acid, and some amide groups. Methanol and acetic acid are relevant in the esterification of galacturonic acid residues for developing the inherent structure functionalities of pectin. The degree of methylation (DM) is a helpful tool for describing the structure of pectin and potential applications; high methoxy pectins (HM) contain more than 50% of carboxyl groups in methylated form, while those with lower content are defined as low methoxy pectins (LM). Most common native pectins are characterized by being methyl esterified. Likewise, acetylation in pectins rarely occurs in native pectins. The degree of acetylation (DA) in pectins is defined as the percentage of galacturonosyl residues that can be acetylated per unit of monosaccharide. DA can be larger than 100% and is usually found in the branched RG regions. In pectins from citrus and apple, the acetyl groups in the HG region are present in low content, rather than in pectins from sugar beet and potato, where higher amounts have been found [13, 14]. Amidation of pectins does not occur naturally; instead, it is induced chemically or enzymatically to improve the functional properties such as solubility in water, gelling, and rheological properties through modifying some non-esterified carboxyl groups into amide groups by using various amino compounds [15–17].

3. Conventional sources of pectin and their applications

Commercial pectin is generally obtained from citrus peels (25% dry matter) and apple pomace (15–18% dry matter), their processing subproducts, and sugar beet pulp (25% dry matter) [18]. The most significant part of commercial pectins includes 85.5% from citrus peels, 14% from apple pomace, and ~ 0.5% from sugar beet pulp [19]. Industrial processes for the extraction of pectin are based on the thermal hydrolysis of the citric peels (mainly from orange, lemon, and lime), apple pomace, and sugar beet pulp by using hot mineral acids like HCl, H_2SO_4, or HNO_3 (~pH 1.5) at ~85°C [20], where the control of the extraction conditions is of great relevance for minimizing the de-esterification and depolymerization of the polysaccharide and improving the functional properties of pectins as gelling, fiber enrichment, stabilizer, texture, and rheology control agent [21]. Notably, these pectin extraction processes generate large amounts of acidic industrial wastes and high energy consumption [22]. Hence, recent investigations have explored the use of more green technologies to overcome these environmental issues and enhance yield extraction [23]. **Table 1** shows some physicochemical properties of pectins obtained from conventional sources using different extraction procedures.

Source		Extraction conditions and yield	Functional Properties	References
Citrus peel	Lime	HHP extraction Enzyme treatment pH: 4.5; 50°C, 4 h. Yield: 26.1%	DE: 75.7%; GalA: 82.8% MW:308.4 kDa; η_{int} 5.0 dL/g Enhanced solubilization Shorter gel setting time	[23]
		Enzyme treatment pH 3.5; 50 °C, 4 h. Yield: 23%	DE: > 82%; GalA:81–84% MW: 69 kDa; Gelling properties	[24]
	Orange	USE: 150 W, 20 kHz, 10 min, 50 °C Citric acid pH 1.5 Yields: 28.1%	GalA: 72%; DE: 37.5% Surface tension: 42.1–46.6 mN/m (0.1–0.5% wt.) WHC: 3.10 gwater/gpectin OHC:1.32 goil/gpectin	[25]
	Pomelo	Conventional extraction Nitric acid pH 2.0; 90 °C; 1.5 h Yield: 23.19%	DE: 57.9%; MW: 353 kDa Viscoelastic solution (<1% wt.) Weak gelation (<1% wt.) Newtonian behavior (<0.4% wt.) Pseudoplastic behavior (>0.4% wt.)	[26]
	Mandarin	HHP: 500 MPa, 10 min Citric acid pH 1.4 Yield: 21.95%	GalA: 75.4–84.4%; DE: 67.7–70.4% MW:1201–2626 kDa Pseudoplastic behavior (3% wt.)	[27, 28]
	Grapefruit	USE:800 W;20 kHz; 58% amplitude 67.8 °C; 30 min Yield: 27.3%	DE: 65.5%; GalA: 50% η_{int}: 3.26 dL/g; MW: 109.5 kDa	[29, 30]
		Conventional extraction H2SO4 pH; 80 °C 1 h. Yield:33.6%	DE: 71.7%; η_{int}: 18.36 dL/g MW 2.3x105 kDa	[31]
Apple pomace	Granny Smith; Royal; and Golden varieties Pomace used for commercial pectin	Chemical or enzyme treatment Yield:4.2–19.8%*	GalA:18.0–67.9%*; DE:52.5–76.4% * DM:58–88%; MW:68–790 kDa *	[32–34]
Sugar beet pulp		Conventional extraction HCl, pH 1.2, 90 °C, 3 h.	DE: 38.6–40.8% Pseudoplastic behavior (2%wt.) Emulsifying activity	[35, 36]

*HHP: High hydrostatic pressure; MW: Molecular weight; η_{int}: Intrinsic viscosity; USE: Ultrasound-assisted extraction; WHC: Water holding capacity; OHC: Oil holding capacity. *Depending on the variety of pomace.*

Table 1.
Physicochemical properties of some pectin extracted from conventional sources.

Pectin is widely used in the food industry as an excellent thickener agent for producing jellies and jams, a pH stabilizer in dairy products and low-calorie products, and an emulsifier in pharmaceutics for the design of drugs to treat gastrointestinal disorders, blood cholesterol reduction, and cancer treatment as well as good former of edible films and coatings, foams, and paper substitutes [17, 24]. Because of the

DOI: http://dx.doi.org/10.5772/intechopen.109579

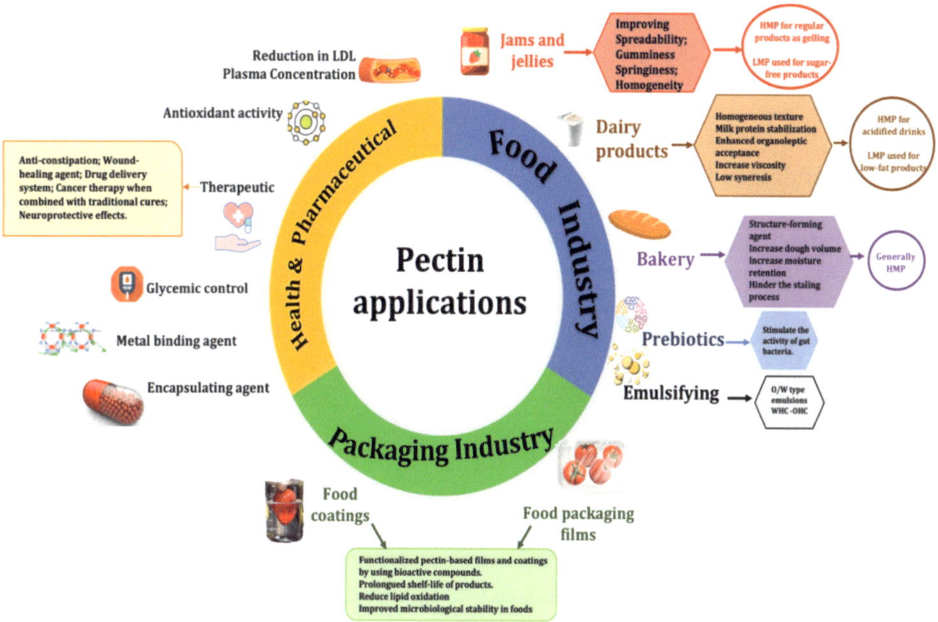

Figure 2.
Principal applications of commercial pectin in food, packaging, and pharmaceutical industries [17].

functional properties of pectins, both LM and HM, many applications in food, industrial, and pharmaceutic sectors can be considered (**Figure 2**).

Most commercial pectins are facilitated to dissolution when a dextrose content is present. An additional pectin classification is based on its gelling capacity, which is relevant in product processing and preservation. Pectins are classified as rapid-set pectin, when gels are formed, preferably at high temperatures, generally used for jams because it reduces the possibility that the fruit rises to the surface before the pectin gel is set, and slow-set pectin, which is preferred in jellies because it allows handling the product before the gel setting without damaging the texture and firmness of the product [25].

Despite the presence of extensive contents of pectin polysaccharide in fruit subproducts, like citrus peels, apple pomace, or sugar beet pulp, it is not the most critical parameter to define a lucrative extraction and the best functional properties for this functional agent [17]; the exploration of novel sources of pectins is raising the attention of scientists and technologists.

4. Vegetable sources of pectin

New sources of pectin that are receiving significant interest in the scientific field are those obtained from different kinds of vegetables, such as pumpkin, eggplant, chayote, and *Opuntia ficus* indica cladodes.

Several studies have been conducted on the extraction of pumpkin pectin using different extraction methods, such as the chemical acid treatment (0.1 M HCl) or enzymatic extraction, where the last has given much higher yields than the acid extraction [26]. Pumpkin pectin fraction **A** was obtained from raw pumpkin with an enzyme preparation of cellulase and α-amylase. Pumpkin pectin fraction **B** was obtained by treating the solution of fraction **A** with pronase to reduce the protein content. The pumpkin pectin fractions **A** and **B** yielded 10.03 and 8.08 g/100 g, respectively.

The DE values of about 47% for pumpkin pectin fractions **A** and **B** were not significantly different, while the GalA contents represent 75.02 and 78.22 g/100 g, respectively. This finding indicated that both fractions are mainly composed of HG [27]. Small amounts (about 10 g/100 g) of six different neutral sugars were found in both pectin fractions, including rhamnose, arabinose, galactose, glucose, xylose, and mannose.

FT-IR and 1D NMR analyses revealed that the pumpkin pectin backbone is mainly composed of 1,4-D galacturonic acid, in which a considerable portion of galacturonic acid residues is present as methyl esters, and L-rhamnose is involved in the linear region of the backbone through α-1,2 linkages. The emulsifying capacity and stability of pumpkin pectin fraction **A** were 63.7 and 58.3%, respectively. At the same time, both properties were not detected in pumpkin pectin fraction **B**. Pectin fraction **A** exhibited emulsifying properties in the water–oil mixture, evidencing the presence of hydrophobic protein components in the pectin structure. In contrast, protein removal in fraction **B** resulted in a loss of emulsifying properties [26]. Therefore, pumpkin pectin could be used as an emulsifying agent in the preparation of oil-in-water emulsions for the beverage industry as long as residual hydrophobic protein components are not removed.

Eggplant fruit (*Solanum melongena* L.), a popular vegetable with an elongated oval shape and dark purple peels, grows worldwide, especially in tropical and subtropic regions. Under optimal extraction conditions by the ultrasound-assisted extraction method (UAE) (ultrasound power of 50 W, irradiation time of 30 min, and pH of 1.5), the pectin extracted from the peels of this vegetable (EPP) indicated that the EPP had a high GalA content (66.08 g/100 g) [28]. Considering the Food and Agriculture Organization (FAO) and European Union recommendations, the GalA content of pectin used as a food additive or pharmaceutical purpose should not be lower than 65 g/100 g pectin. This pectin had a high DE (61.22%) and was categorized as HM pectin (DE > 50%). EPP had a protein content of 2.53 g/100 g, which can be attributed to the difference in raw materials and extraction techniques. However, FAO suggests that the protein content of pectin should not be higher than 15.6 g/100 g [24]. In addition, EPP showed good values in functional features such as water-holding capacity (WHC) and oil-holding capacity (OHC). Under the optimal extraction conditions, EPP exhibited a WHC of 6.22 ± 0.21 g water per g EPP, while the OHC was 2.12 ± 0.15 g oil per g EPP. The emulsifying activity (EA) and emulsifying stability (ES) of EPP were evaluated, EA was about 56.16%, and the highest emulsion stability was 96.36 ± 0.80 at 4°C. EPP also exhibited antioxidant activity, determined by the DPPH radical scavenging method, reaching a highest antioxidant activity at a concentration of 50 mg/mL (94%), which was still lower than the antioxidant activity performed by the ascorbic acid, with an IC_{50} value of 1.39 mg/mL; this activity is due to the higher total phenolic content (TPC = 96.81 ± 2.18 mg GAEa/g pectin) associated to the EPP. The GalA content of the extracted pectin can be also effective in the antioxidant activity due to active portions in its structure [29].

Chayote is one of the most cultivated vegetables in the world. The major producing countries are Mexico, Brazil, and China [30]. The UAE method has been used to extract chayote pectin (PEUO) [31]. Using a liquid/solid ratio of 50 mL/g, a temperature of 70°C, and an ultrasonic time of 40 min as optimal extraction conditions. The yield was around 6.19%. Under these extraction conditions, PEUO exhibited a low DE (17.6%), indicating that the chayote pectin could be considered as LM pectin.

This property could be attributed to the harsh extraction conditions that would promote the de-esterification of polygalacturonic chains. The GalA content in PEUO accounted for 57.25%. To our knowledge, the ripeness, blanching, ultrasound, and other effects may influence the GalA content in the extracted pectin [31], besides the contribution to improve the depolymerization of polysaccharides, releasing the water-soluble pectin from the plant tissue [32]. The molecular weight in pectins significantly affects the emulsification, rheology, and their colloid stability. In this sense, the weight-average molecular weight and number-average molecular weight of PEUO were 2.47×10^6 g/mol and 1.29×10^6 g/mol, respectively, and the polydispersity index was 1.91. Polydispersity index higher than 1 suggests that PEUO extracted by UAE represents a heterogeneous natural polysaccharide with a broad range of polymer size distribution [31]. The monosaccharide composition of PEUO indicated the presence of five monosaccharides, where glucose (Glu) represents the most abundant monosaccharide (90.6%), followed by Gal (8%), D-Xyl (0.6%), Ara (0.6%), and Rha (0.2%). Besides, the content of Gal was significantly higher than that of Ara, indicating that the RG-I region may have been highly branched with galactan or arabinogalactan. Rheological properties of PEUO aqueous dispersions (<5%wt.) exhibited a non-Newtonian behavior [31]. Other functional properties like WHC and OHC for PEUO showed suitable values for both WHC (3.14 ± 0.42 g water/g PEUO) and OHC (3.73 ± 0.30 g oil/g PEUO). High WHC in PEUO makes it suitable as a food industry thickener. EA and ES were determined at 4°C and 25°C. The ES for PEUO emulsions were 88.36 ± 5.63% and 81.28 ± 4.82% after 1 day, and these values changed after 30 days to 85.33 ± 4.16% and 77.59 ± 5.19%, respectively. The lower temperature (4°C) was presumably more suitable for storing the PEUO emulsion. These results provide further evidence that chayote pectin may have great potential to be applied as an emulsifier and stabilizer in the food industry [31, 33]. Regarding the antioxidant activity of PEUO, it was higher when compared to pectin extracted from apples. Due to its techno-functional properties, PEUO may be used as a gelling agent and preservative in jam production or as a viscosity enhancer in beverages.

Another source of pectin that has received much attention is the *Opuntia ficus* indica (OFI) cladodes. This pectin has been extracted by acid water, ultrasound, and enzyme treatments [34, 35]. The pectin obtained by ultrasound under optimal conditions (sonication time of 70 min, temperature of 70°C, pH of 1.5, and water:solid ratio of 30 mL/g) reached an extraction yield of 18.14% ± 1.41%, with a GalA content of 68.87%. This pectin had a DE of 41.42%, classifying it as an LM pectin [36]. This DE value was higher than that achieved when the OFI pectin was extracted by the chemical process, which was 30.67% [37]. WHC in OFI pectin was 4.84 g water/g OFI pectin, ultrasound-induced cavitations in the pectin structure improving the water penetration and its absorption [38]. WHC for OFI pectin extracted by the chemical process was higher (5.64 g water/g OFI pectin) [34]. OHC for OFI pectin extracted with ultrasound was 1.01 g oil/g OFI pectin, slightly lower than pectin extracted by the chemical method (1.24 g oil/g OFI pectin) [34]. EA and ES were determined at two pectin concentrations (2 and 4% w/v). EA values were 19.23% and 26.92%, respectively, showing that the emulsion stability depends on the pectin concentration. OFI pectin at 4% maintained stability of more than 57% of the emulsion after 30 min of incubation at 80°C, unlike the 2% pectin solution, which could not retain more than 40% of the emulsion. This stability of the emulsions could be attributed to the rise of viscosities in the pectin solutions caused by the formation of a layer of pectin around each oil droplet, delaying the coalescence phenomenon [39, 40]. This stability was affected by the high pectin extraction temperature (> 45°C) [41]. ES in

OFI pectin extracted with acid water at 2% displayed higher values (90.45%) [34]. Differences in the ES are due to differences in the extraction methods, which affect the average molecular weight and the GalA content in the structure of pectin and therefore influencing the long-term stability in the emulsions [42]. When enzyme treatments were used for OFI pectin extraction, the optimal conditions were cellulase/xylanase at an LS ratio of 22 mL/g, cellulase/xylanase ratio of 2 U/U, and enzymes/matter ratio of 4 U/g, reaching an extraction yield of 17.91% [35], being more effective than the chemical treatment, which resulted in an extraction yield of 6.13 ± 0.60% [34]. Enzyme-assisted extraction of pectin depends on the choice of enzymatic activities based on the strength of pectin connection with cellulose and xylan and their abundance in the cell wall of the plant source [43].

For OFI pectin, the total sugar content was 89.94%, the main monosaccharide was GalA (66.66 ± 2.46%), with a DE of 35.04%, which was higher than that reported by Lira-Ortiz et al. [44] for pectin from prickly pear fruits (*Opuntia albicarpa*; DE 30.7%). OFI pectin had a WHC of 5.42 ± 0.16 g water /g OFI pectin, slightly lower than that for pectin extracted by the chemical process (5.64 g water/g OFI pectin) [34]. Various intrinsic factors, like the chemical structure of the biomaterial, and extrinsic factors, such as the pH, temperature, and ionic strength, can affect the WHC [45]. The OHC value of pectins was 1.23 ± 0.42 g oil/g OFI pectin. It was like the OHC of the OFI pectin extracted by the acid water method [34]. Thus, the oil retention power depends essentially on the hydrophilic nature and the overall charge density of the constituents [45]. EA values for OFI pectin emulsions at 2 and 4% were 26.9% and 30.77%, respectively. These values were lower than the ones found by Bayar et al. [34] for a 2% concentration of pectin extracted by the chemical process from the OFI cladodes (35%), proving that the extraction process influences the functional properties of pectin macromolecules [46]. The ES rates were 14.31% and 87.48% for 2% and 4% of pectin hydrocolloid in the emulsions, this long-term stability when emulsions were submitted to temperature treatment at 80°C is due to the high viscosity of pectin solution and by the formation of layers around the fat globules by the pectin [39].

5. Unconventional sources of pectin: hulls or husks and seeds

It is well known that the primary sources of pectin extraction are those obtained from citrus fruits or apples, due to their high yield and physicochemical properties that make them useful for various applications in the food and pharmaceutical industries. However, in recent years, new extraction sources have been sought that may represent alternatives to overexploited sources and that also have the advantage of allowing the use of organic by-products, such as the case of hulls or husks and seeds, from which pectins with specific physicochemical properties of high utility for multiple applications can be obtained.

Table 2 shows current research work regarding unconventional sources for obtaining pectins, classified as hulls or husks that come from dry fruits (almonds, pistachios, walnuts, and cocoa), pods, and legume seeds (soy, peas, faba beans, and riang), cereal leaves (*Zea mays*) and seeds of different fruits (*Nicandra physaloides* Linn., Gaertn, papaya, jackfruit, creeping fig and sesame). In addition, its extraction methods and its most outstanding properties are also described.

The most widely used pectin extraction method for hulls or husk and seeds is the conventional one, which consists of acidifying the sample, for which different types of organic (citric and oxalic acid) and inorganic (HCl and HNO_3) acids are used; the type of acid used influences the extraction conditions and the properties of the pectin obtained [30]. Subsequently, a heat treatment is carried out using a conventional hot plate, or, for more efficient extraction, it can be assisted by microwaves [49] or

Pectin source	Extraction conditions and Yield	Functional properties	Reference
Hulls or husks			
Almond hull	First part: acidification with citric acid of almond hull pectin whose optimal conditions were pH = 1.4, liquid-solid ratio (LSR) 20.13, 90°C for 58.65 min followed by filtration. Second part: a mixture of pectin supernatant with 96% ethanol at a ratio of 1:1 v/v, then the precipitate obtained was dried in an oven at 50°C. Yield: 26.32% wt.	Extraction of LM pectin DE: 26.4% Forming forms gels using Ca^{2+} at a pH 3–7 Do not need sugar to form gel High polydispersity due to the small chains formed during the extraction process.	[28]
Pistachio hull	Conventional and ultrasound-assisted, Acidification with a citric acid solution Yield: 32.3%.	Extraction of LM pectin DE = 19.29% Maximum emulsifying capacity with 6% wt. pectin, EA index: 172.85 ± 0.59 m^2/g ES index: 158.28 ± 3.41 min, High creaming stability Shear-thinning behavior.	[47]
A green husk of walnuts (*Juglans regia* L.)	Walnuts husks from different regions of cultivation. Walnut husks powder heated in an acid medium. Ethanol precipitation. The pectin was decolorized using acetone.	The soil and climate conditions where the walnut husks were obtained caused variations in the properties of the pectins obtained. DE: higher 65%. Lamellar and leaf-shaped structures, depending on the region of cultivation	[48]
Cocoa pod husk (*Theobroma cacao*)	Microwave-assisted extraction using an acidified medium with oxalic acid. pH 1.16 15 min. Liquid/solid ratio:25 Yield: 9.64%.	The decrease in pH during extraction produced a decrease in the esterification degree which reduced the gelling ability of pectin. This could be observed by FTIR spectroscopy.	[49]
Legumes (from seeds or pods)			
Soy hull	Thermal treatment by microwave irradiation Acidification 0.6% wt. citric acid or sodium citrate (SC). Pectin precipitation with ethanol.	The pectin extracted with SC had better stability of emulsions, smaller droplet sizes and greater emulsifying capacity. Applied into mayonnaise, achieving uniformly distributed drops and high stability.	[50]
Soy hull	Pectin extraction was carried out from milled soy hulls, from which galactomannan was removed. Acid medium, HCl followed by HNO_3.	Low uronic acid content. Low yield. Xylogalacturonan and rhamnogalacturonans as major components. Cannot form gels by adding Ca^{2+}.	[51]

Pectin source	Extraction conditions and Yield	Functional properties	Reference
Pea hull	To optimize the extraction, a central composite design was carried out where the effect of pH, temperature, and time on the yield and purity of the pectins was evaluated using two different acid media: citric acid and HNO_3. pH 2.0 Yield: 3.5–9.8% with citric acid Yield: 1.4–8.0% with HNO_3. Purity: >65%, related to the high uric acid content.	LM pectin Mainly composed of xylogalacturonan.	[52]
Faba bean hull	Microwave assisted extraction (640 W). pH = 1.5 with 1 M HCl, 9 min. Ethanol precipitation. Yield: 14.86%	HM pectin DE: 54.08%.	[53]
Riang (*Parkia timoriana* (DC.) Merr.) pod husk	Acid water pH = 2 with HNO_3, heated to 90°C for 90 min. Final pH adjusted to 4.5. Yield: 15.0%	HM pectin DE: ~66%. Pseudoplastic behavior at a concentration > 2%w/v, Newtonian behavior at concentrations <2% w/v. High antioxidant activity and high content of phenolic compounds (mainly tannins).	[54]
Cereal			
Zea mays husk	Extraction with high-power ultrasound (US) application. Pretreatment in a primary medium, followed by enzymatic hydrolysis with cellulase (pH = 5.2), precipitation of pectin with ethanol, and subsequent lyophilization to obtain the MP fraction. Second treatment consisted of high-power ultrasound at 20 kHz, plus the steps of the first treatment to obtain the MP-US fraction.	Formation of thermo-irreversible and soft gels at pH = 6 in the presence of Ca^{2+} Interaction with the iron (II) ion to form thermo-reversible weak gels. LM pectin and high-water solubility due to the ultrasound treatment.	[55]
Fruit seeds			
Papaya seeds	Acid medium (citric acid) 80°C, pH 1.5, 60 min Yield: 8.66%	LM: 9.22%.	[56]
Jackfruit seeds sheats	Pectin extraction was from the slimy sheats of the jackfruit seed (JS). Water acidified with oxalic acid 90°C, 1 h. Yield: 35.52%	Total phenolic content: 65.7 mg GAE/g Antioxidant activity: DPPH method: 25.29 ± 4.03% FRAP: 10.4 μM	[57]

Pectin source	Extraction conditions and Yield	Functional properties	Reference
Creeping fig fruit seeds	Chemical extraction in acid conditions.	LM pectin DE: ~20%. Form gels at low pH values with the addition of glucono-δ-lactone Viscous solutions are obtained at pH = 4.5 with Na^+ and K^+ ions for favoring the formation of the gel. Gel strength depends on the type of salt added and its concentration.	[58]
Nicandra physaloides (Linn.) Gaertn seeds	Enzyme inactivation of the NPG seeds were inactivated with heating. Aqueous extraction at 60°C. Different fractions of pectin were obtained. Yield: 9.17–10.56%	LM pectin DE = 46.93%. Spontaneous gel formation at 1.5%. Gel formation at <1.5% in the presence of NaCl and KCl.	[59]
Sesame seed hull	Defatted seed Acid medium (HCl). Fractionation with ethanol (30%, 50%, and 90%). Maximum yield: 75.6% at 30% ethanol	High antioxidant activity Able to stabilize emulsions.	[60]

Table 2.
Unconventional sources of pectins: Hulls or husks and seeds.

by high-power ultrasound [61]. Finally, separation is carried out using the ethanol solvent, and the pectin obtained is dried.

Among the properties shared by pectins obtained from hulls or husks and seeds is that they are primarily LM pectins, with very varied DEs, and they also can form gels in the presence of ions such as calcium, sodium, or potassium [28]. However, the esterification degree influences the properties of the gels formed [49]. Besides, this type of pectin regularly achieves the formation of stable emulsions [24, 62], a shear-thinning rheological behavior and can even, in some cases, present antioxidant activity [14, 54, 63].

One of the main disadvantages of obtaining pectin from hulls or husks and seeds is its low yield (regularly less than 15%), since the extraction is carried out by conventional methods, where conditions such as the type of acid influence the yield obtained. However, research is currently being carried out on new methods that allow a more efficient extraction and higher yield to encourage the use of unconventional sources of pectin.

6. Applications of new sources of pectin

Nowadays, green chemistry leads to environmentally friendly bioproduct extraction approaches. Because bioproducts are biocompatible, they have a wide range of applications [33]. The synthesis and production of bioproducts use substantially less energy and solvent, and they can now be scaled up with a small initial expenditure [14, 40]. Biomaterial formulations, sometimes inspired by biomimicking nature's behavior, are specifically tailored for applications involving human consumption products and innovative biobased materials [64]. In the field of biomaterials, hydrogels have gained popularity owing to their specific properties, such as biodegradability, biocompatibility, a soft-wet feel, and resemblance to organic tissue. Hydrogels with tridimensional crosslinked polymeric structures made from natural polymers have been extensively studied because of the increasing need for biomaterials with novel features for human consumption-related applications [65]. Pectin, a biopolymer found in the cell walls of fruits and vegetables, is extensively employed in the food, pharmaceutical, and textile sectors due to its ability to produce a thick gel-like solution [65]. Pectin is a gelling ingredient in the production of jams, jellies, and marmalades. Over the past decade, intense new research has yielded a new understanding of its molecular structure and physiological function, opening the gate to novel manufacturing techniques and entirely new applications, such as new advanced biomaterials, for example, calcium phosphate pectin for bone restoration and biobased construction, and building materials, for example, pectin aerogels for thermal insulation [64, 66].

According to the scientific literature, we can classify applications of pectins in two ways: first, according to their physicochemical properties, and last, according to their field of application. The specific application of each of the novel pectin sources is intimately linked to their particular physicochemical characteristics; please see **Figure 3**. For example, LM pectin is believed to be a helpful stabilizer for dairy products. This is due to low methoxyl pectin gels in the presence of divalent cations, in this specific instance, calcium ions. The capacity of HM pectin to gel at moderately lower pH values (pH 2–3.5) in the addition of soluble substances, such as sucrose, makes it suitable for use in the preparation of jams and jellies [67]. An LM pectin is an attractive option for use as a gelling agent in manufacturing low-calorie jams due to its ability to form a gel without added sugar. Unlike gums, which impart a slimy mouth feel, the use of pectin to increase the viscosity in soft drinks and beverages gives a clean mouth feel;

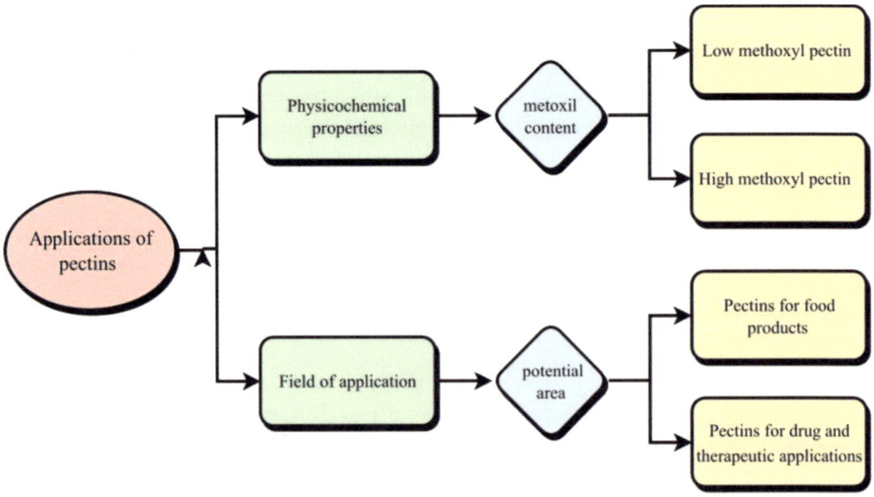

Figure 3.
Different classifications for pectin.

Application		Source	Extraction technique	Reference
Pectins for food products	Stabilizer for dairy products	Grapefruit peel	Acid hydrolysis	[68]
	Food films, gelling agents, and plasticizer	Lime peel	Citric acid-microwave extraction	[69]
	Food packaging	Lemon waste peel	Microwave extraction	[70]
	Films and emulsions	Citrus	—	[71]
	Antioxidants in food formulations	Jackfruit peel	Ultrasonic-microwave extraction	[72]
Pectins for drug and therapeutic applications	Drug delivery systems	Citrus	—	[73]
	Drug delivery systems	Fig skin	Ultrasonic-microwave extraction	[74]
	Tissue engineering	Lemon peels	Acid hydrolysis	[75]
	Bioprinting of 3D scaffolds	Citrus peels	—	[76]
	Wound healing	*Akebia trifoliata* fruit peel	Acid hydrolysis	[77]
	Accelerated wound healing	Cyclea Barbata Miers	Cold acid hydrolysis	[78]
	Skin wound healing	Papaya fruit	—	[79]

Table 3.
Applications of novel pectins.

this may be due to the low viscosity of low-concentration pectin solutions at the shear rate of the mouth [67].

From the viewpoint related to their field of application, pectins may be categorized into pectins for food products and pectins for drug and therapeutic applications. Please see **Table 3**. Current research trends in food packaging promote the development of biodegradable, renewable, and environmentally friendly materials. Pectin-based edible coatings are among the most recent advancements in the world of food packaging. Including additional biopolymers, such as cellulose and natural compounds with antioxidant and antibacterial properties, has enhanced and strengthened these coatings.

Additionally, researchers have discovered the biological functions of pectin, consequently increasing its application in the pharmaceutical industry, including drug delivery systems, skin and bone tissue engineering, and wound dressings [65, 76]. Pectin is most widely used in the formulation of drugs for oral administration, such as tablets, gels, hydrogels, beads, aerogels, and coated and compression-coated doses. The ability of pectin to withstand acidic conditions and higher temperatures allows for the development of drug delivery systems able to load and release drugs at a specific location. Pectin has primarily been considered a colon-specific drug delivery vehicle that reduces systemic toxicity while increasing bioactivity and medication stability.

Pectin also has significant potential for use in tissue engineering. Pectins may promote mineral nucleation in this application if immersed in the appropriate physiological conditions, resulting in biomimetic structures that more closely resemble the natural architecture of bone. Furthermore, pectins are responsible for wound healing treatments' gelling protection and anti-inflammatory effects [76]. By crosslinking pectins, calcium ions aid in its gelation. Solubilized pectin forms an acidic environment that acts as a bacterial or viral barrier, and pectin hydrogels allow for the loading and release of drugs such as antibiotics, analgesics, and tissue repair agents. Other physiological effects of pectin have been described, such as prebiotic, antimicrobial, antiglycation, and antioxidant. Pectin has also been used to nano-encapsulate bioactive substances, thereby increasing their shelf life and stability.

The exploration of new sources of pectin, involving the introduction of cleaner and new sustainable extraction techniques, demands more research to guarantee that an industrial application is sustainable and competitive in the current market.

7. Conclusions

Pectin is one of the primary polysaccharides present in plants; it contributes to the physical and nutritional value of plant-based goods. It's a macromolecule that can create flexible polymer chains. Source and extraction circumstances affect its functioning characteristics. Citric fruits and apples are the principal sources of commercial pectin, although non-conventional sources have been examined, including agro-industrial sub-products and wastes, pulps, husks, hulls, peels, Cactaceae, and vegetables. Pectin has been functionalized by chemical or enzyme processes that affect its physical characteristics, such as molecular weight, degree of esterification (DE), and surface charge, leading to new functional or enhanced qualities as well as new techniques and applications. Pumpkin, eggplant, chayote, and *Opuntia ficus* indica cladodes are new sources of pectin. Due to their high production and physicochemical qualities, citrus fruits and apples are the principal sources of pectin extraction. In recent years, new extraction sources have been sought that may represent alternatives to overexploited sources and that allow the use of organic by-products, such as hulls or husks and seeds, from which pectin with specific physicochemical properties can be obtained for multiple applications. Intense new research has yielded a new understanding of its molecular structure and physiological function, opening the door to novel manufacturing techniques and entirely new applications, such as calcium phosphate pectin for bone restoration and pectin aerogels for thermal insulation.

Acknowledgements

The authors wish to acknowledge the partial financial support of this research to the Universidad Autónoma del Estado de México through project 6661/2022SF.

Conflict of interest

The authors declare no conflict of interest.

Author details

Stefani Cortés-Camargo[1], Angélica Román-Guerrero[2], Erik Alpizar-Reyes[3] and César Pérez-Alonso[4]*

1 Nanotechnology Department, Technological University of Zinacantepec, Zinacantepec, Estado de México, México

2 Biotechnology Department, Metropolitan Autonomous University-Iztapalapa, CDMX, México

3 Department of Civil and Environmental Engineering, LabMAT, Bio Bio University, Concepción, Chile

4 Faculty of Chemistry, Department of Chemical Engineering, Autonomous Mexico State University, Toluca, Estado de México, México

*Address all correspondence to: cpereza@uaemex.mx

References

[1] 2023 Pectin Market Outlook Report - Market Size, Market Split, Market Shares Data, Insights, Trends, Opportunities, Companies: Growth Forecasts by Product Type, Application, and Region from 2022 to 2030. https://www.researchandmarkets.com/reports/5686321/2023-pectin-market-outlook-report-market-size#product--description [Accessed: November 21, 2022)

[2] Freitas CMP, Coimbra JSR, Souza VGL, et al. Structure and applications of pectin in food, biomedical, and pharmaceutical industry: A review. Coatings. 2021;**11**:922. DOI: 10.3390/COATINGS11080922

[3] Herrera-Rodríguez SE, Pacheco N, Ayora-Talavera T, et al. Advances in the green extraction methods and pharmaceutical applications of bioactive pectins from unconventional sources: A review. Studies in Natural Products Chemistry. 2022;**73**:221-264. DOI: 10.1016/B978-0-323-91097-2.00015-7

[4] Muñoz-Almagro N, Montilla A, Moreno FJ, et al. Modification of citrus and apple pectin by power ultrasound: Effects of acid and enzymatic treatment. Ultrasonics Sonochemistry. 2017;**38**:807-819. DOI: 10.1016/J.ULTSONCH.2016.11.039

[5] Muñoz-Almagro N, Montilla A, Villamiel M. Role of pectin in the current trends towards low-glycaemic food consumption. Food Research International (Ottawa, Ont). 2021;**140**. Epub ahead of print 1 February. DOI: 10.1016/J.FOODRES.2020.109851

[6] Zhang H, Chen J, Li J, et al. Extraction and characterization of RG-I enriched pectic polysaccharides from mandarin citrus peel. Food Hydrocolloids. 2018;**79**:579-586. DOI: 10.1016/j.foodhyd.2017.12.002

[7] Willats WGT, Knox JP, Mikkelsen JD. Pectin: New insights into an old polymer are starting to gel. Trends in Food Science & Technology. 2006;**17**:97-104. DOI: 10.1016/J.TIFS.2005.10.008

[8] Leclere L, Van CP, Michiels C. Anti-cancer activities of pH- or heat-modified pectin. Frontiers in Pharmacology. Oct 2013;**4**:128. DOI: 10.3389/FPHAR.2013.00128/BIBTEX

[9] Atmodjo MA, Hao Z, Mohnen D. Evolving views of pectin biosynthesis. Annual Review of Plant Biology. 2013;**64**:747-779. DOI: 10.1146/annurev-arplant-042811-105534

[10] Ndeh D, Rogowski A, Cartmell A, et al. Complex pectin metabolism by gut bacteria reveals novel catalytic functions. Nature. 2017;**544**(7648):65-70. DOI: 10.1038/nature21725

[11] O'Neill MA, Ishii T, Albersheim P, et al. Rhamnogalacturonan II: Structure and function of a borate cross-linked cell wall pectic polysaccharide. Annual Review of Plant Biology. 2004;**55**:109-139. DOI: 10.1146/ANNUREV.ARPLANT.55.031903.141750

[12] Dranca F, Oroian M. Extraction, purification and characterization of pectin from alternative sources with potential technological applications. Food Research International. 2018;**113**:327-350. DOI: 10.1016/j.foodres.2018.06.065

[13] Ngenefeme FTJ, Eko NJ, Mbom YD, et al. A one pot green synthesis and

characterisation of iron oxide-pectin hybrid nanocomposite. Open Journal of Composite Materials. 2013;**03**:30-37. DOI: 10.4236/ojcm.2013.32005

[14] Alpizar-Reyes E, Cortés-Camargo S, Román-Guerrero A, et al. Chapter 14: Tamarind gum as a wall material in the microencapsulation of drugs and natural products. In: Micro- and Nanoengineered Gum-Based Biomaterials for Drug Delivery and Biomedical Applications. 2022. pp. 347-382. DOI: 10.1016/B978-0-323-90986-0.00016-9

[15] Chen J, Niu X, Dai T, et al. Amino acid-amidated pectin: Preparation and characterization. Food Chemistry. 2020;**309**. Epub ahead of print 30 March. DOI: 10.1016/J.FOODCHEM.2019.125768

[16] Zheng C, Zou Y, Huang Y, et al. Biosynthesis of amidated pectins with ultra-high viscosity and low gelation restriction through ultra-low temperature enzymatic method. Food Hydrocolloids. 2023;**134**. Epub ahead of print 1 January. DOI: 10.1016/J.FOODHYD.2022.108037

[17] Chandel V, Biswas D, Roy S, et al. Current advancements in pectin: Extraction. Properties and Multifunctional Applications. Foods. 2022;**11**:1-30. DOI: 10.3390/foods11172683

[18] Mesbahi G, Jamalian J, Farahnaky A. A comparative study on functional properties of beet and citrus pectins in food systems. Food Hydrocolloids. 2005;**19**:731-738. DOI: 10.1016/J.FOODHYD.2004.08.002

[19] Ciriminna R, Chavarría-Hernández N, Inés Rodríguez Hernández A, et al. Pectin: A new perspective from the biorefinery standpoint. Biofuels, Bioproducts and Biorefining. 2015;**9**:368-377. DOI: 10.1002/BBB.1551

[20] Chen J, Cheng H, Zhi Z, et al. Extraction temperature is a decisive factor for the properties of pectin. Food Hydrocolloids. 2021;**112**:106160. DOI: 10.1016/J.FOODHYD.2020.106160

[21] Ciriminna R, Fidalgo A, Delisi R, et al. Controlling the degree of esterification of citrus pectin for demanding applications by selection of the source. ACS Omega. 2017;**2**:7991-7995. DOI: 10.1021/ACSOMEGA.7B01109/ASSET/IMAGES/AO-2017-01109E_M002.GIF

[22] Panouillé M, Thibault JF, Bonnin E. Cellulase and protease preparations can extract Pectins from various plant byproducts. Journal of Agricultural and Food Chemistry. 2006;**54**:8926-8935. DOI: 10.1021/JF0617824

[23] Munarin F, Tanzi MC, Petrini P. Advances in biomedical applications of pectin gels. International journal of biological macromolecules. 2012;**51**:681-689. DOI: 10.1016/J.IJBIOMAC.2012.07.002

[24] Alpizar-Reyes E, Cruz-Olivares J, Cortés-Camargo S, et al. Structural, physicochemical, and emulsifying properties of pectin obtained by aqueous extraction from red pitaya (Hylocereus polyrhizus) peel. Revista Mexicana de Ingeniera Quimica. 2022;**21**:Alim2887–Alim2887. DOI: 10.24275/rmiq/Alim2887

[25] Smith DA. Jams and preserves. Methods of manufacture. In: Caballero B, editor. Encyclopedia of Food Sciences and Nutrition. 2nd ed. USA: Elsevier; 2003. pp. 3409-3415. DOI: 10.1016/B0-12-227055-X/00660-X

[26] Cui SW, Chang YH. Emulsifying and structural properties of pectin enzymatically extracted from pumpkin. LWT - Food Science and Technology. 2014;**58**:396-403. DOI: 10.1016/J.LWT.2014.04.012

[27] Vincken JP, Schols HA, Oomen RJFJ, et al. If Homogalacturonan were a side chain of Rhamnogalacturonan I. implications for Cell Wall architecture. Plant Physiology. 2003;**132**:1781-1789. DOI: 10.1104/PP.103.022350

[28] Hosseini SS, Khodaiyan F, Kazemi M, et al. Optimization and characterization of pectin extracted from sour orange peel by ultrasound assisted method. International Journal of Biological Macromolecules. 2019;**125**:621-629. DOI: 10.1016/J.IJBIOMAC.2018.12.096

[29] Kazemi M, Khodaiyan F, Labbafi M, et al. Pistachio green hull pectin: Optimization of microwave-assisted extraction and evaluation of its physicochemical, structural and functional properties. Food Chemistry. 2019;**271**:663-672. DOI: 10.1016/j.foodchem.2018.07.212

[30] Feliciano GD, Lima EAC, dos Pereira MJS, et al. Effect of a chayotte (Sechium edule) extract on the labeling of red blood cells and plasma proteins with technetium-99m: in vitro and in vivo studies. Cellular and Molecular Biology (Noisy-le-Grand, France). 2002;**48**:751-755

[31] Ke J, Jiang G, Shen G, et al. Optimization, characterization and rheological behavior study of pectin extracted from chayote (Sechium edule) using ultrasound assisted method. International Journal of Biological Macromolecules. 2020;**147**:688-698. DOI: 10.1016/j.ijbiomac.2020.01.055

[32] Su DL, Li PJ, Quek SY, et al. Efficient extraction and characterization of pectin from orange peel by a combined surfactant and microwave assisted process. Food Chemistry. 2019;**286**:1-7. DOI: 10.1016/J.FOODCHEM.2019.01.200

[33] Velázquez-Gutiérrez SK, Alpizar-Reyes E, Cruz-Olivares J, et al. Ionic gelation encapsulation of sesame oil with sodium alginate-nopal mucilage blends: Encapsulation efficiency and oxidative stability. Revista Mexicana de Ingenieria Quimica. 2020;**19**:Alim349–Alim362. DOI: 10.24275/rmiq/Alim1642

[34] Bayar N, Kriaa M, Kammoun R. Extraction and characterization of three polysaccharides extracted from Opuntia ficus indica cladodes. International Journal of Biological Macromolecules. 2016;**92**:441-450. DOI: 10.1016/J.IJBIOMAC.2016.07.042

[35] Bayar N, Friji M, Kammoun R. Optimization of enzymatic extraction of pectin from Opuntia ficus indica cladodes after mucilage removal. Food Chemistry. 2018;**241**:127-134. DOI: 10.1016/j.foodchem.2017.08.051

[36] Bayar N, Bouallegue T, Achour M, et al. Ultrasonic extraction of pectin from Opuntia ficus indica cladodes after mucilage removal: Optimization of experimental conditions and evaluation of chemical and functional properties. Food Chemistry. 2017;**235**:275-282. DOI: 10.1016/j.foodchem.2017.05.029

[37] Forni E, Penci M, Polesello A. A preliminary characterization of some pectins from quince fruit (Cydonia oblonga mill.) and prickly pear (Opuntia ficus indica) peel. Carbohydrate Polymers. 1994;**23**:231-234. DOI: 10.1016/0144-8617(94)90183-X

[38] Chemat F, Rombaut N, Sicaire AG, et al. Ultrasound assisted extraction of food and natural products. Mechanisms, techniques, combinations, protocols and applications. A review. Ultrasonics Sonochemistry. 2017;**34**:540-560. DOI: 10.1016/J.ULTSONCH.2016.06.035

[39] Prajapati VD, Jani GK, Moradiya NG, et al. Pharmaceutical applications of various natural gums, mucilages and

their modified forms. Carbohydrate Polymers. 2013;**92**:1685-1699. DOI: 10.1016/j.carbpol.2012.11.021

[40] Alpizar-Reyes E, Román-Guerrero A, Cortés-Camargo S, et al. Chapter 2: Recent approaches in alginate-based carriers for delivery of therapeutics and biomedicine. In: Polysaccharide-based Biomaterials: Delivery of Therapeutics and Biomedical Applications. 2022. pp. 27-68. DOI: 10.1039/9781839166235-00027

[41] Lv C, Wang Y, Wang LJ, et al. Optimization of production yield and functional properties of pectin extracted from sugar beet pulp. Carbohydrate Polymers. 2013;**95**:233-240. DOI: 10.1016/j.carbpol.2013.02.062

[42] Funami T, Nakauma M, Ishihara S, et al. Structural modifications of sugar beet pectin and the relationship of structure to functionality. Food Hydrocolloids. 2011;**25**:221-229. DOI: 10.1016/J.FOODHYD.2009.11.017

[43] Zykwinska A, Thibault JF, Ralet MC. Competitive binding of pectin and xyloglucan with primary cell wall cellulose. Carbohydrate Polymers. 2008;**74**:957-961. DOI: 10.1016/J.CARBPOL.2008.05.004

[44] Lira-Ortiz AL, Reséndiz-Vega F, Ríos-Leal E, et al. Pectins from waste of prickly pear fruits (Opuntia albicarpa Scheinvar' Reyna'): Chemical and rheological properties. Food Hydrocolloids. 2014;**37**:93-99. DOI: 10.1016/j.foodhyd.2013.10.018

[45] Elleuch M, Bedigian D, Roiseux O, et al. Dietary fibre and fibre-rich by-products of food processing: Characterisation, technological functionality and commercial applications: A review. Food Chemistry. 2011;**124**:411-421. DOI: 10.1016/J.FOODCHEM.2010.06.077

[46] Leroux J, Langendorff V, Schick G, et al. Emulsion stabilizing properties of pectin. Food Hydrocolloids. 2003;**17**:455-462. DOI: 10.1016/S0268-005X(03)00027-4

[47] Barış S, Elik A, Fahrettin G, et al. Pistachio hull as an alternative pectin source: Its extraction and use in oil in water emulsion system. Preparative Biochemistry & Biotechnology. 2022. DOI: 10.1080/10826068.2022.2098323 [Epub ahead of print]

[48] La Torre C, Caputo P, Plastina P, et al. Green husk of walnuts (Juglans regia L.) from southern Italy as a valuable source for the recovery of glucans and Pectins. Fermentation. 2021;7:305. DOI: 10.3390/FERMENTATION7040305

[49] Pangestu R, Amanah S, Juanssilfero AB, et al. Response surface methodology for microwave-assisted extraction of pectin from cocoa pod husk (Theobroma cacao) mediated by oxalic acid. Journal of Food Measurement and Characterization. 2020;**14**:2126-2133. DOI: 10.1007/s11694-020-00459-4

[50] Liu D, Tang W, Huang XJ, et al. Structural characteristic of pectin-glucuronoxylan complex from Dolichos lablab L. hull. Carbohydrate Polymers. 2022;**298**:120023. DOI: 10.1016/J.CARBPOL.2022.120023

[51] Reichembach LH, de Oliveira Petkowicz CL. New findings on acid-extractable pectins from soy hull. Carbohydrate Polymers. 2022;**294**:119831. DOI: 10.1016/J.CARBPOL.2022.119831

[52] Gutöhrlein F, Drusch S, Schalow S. Extraction of low methoxylated pectin from pea hulls via RSM. Food Hydrocolloids. 2020;**102**:105609. DOI: 10.1016/J.FOODHYD.2019.105609

[53] Hussain Dar A, Ul Q, Rizvi EH, et al. Microwave Assisted Extraction of Pectin

from Dried Hull of Faba Bean. www.jpht.in [Accessed: November 21, 2022]

[54] Buathongjan C, Israkarn K, Sangwan W, et al. Studies on chemical composition, rheological and antioxidant properties of pectin isolated from Riang (Parkia timoriana (DC.) Merr.) pod. International Journal of Biological Macromolecules. 2020;**164**:4575-4582. DOI: 10.1016/J.IJBIOMAC.2020.09.079

[55] Higuera-Coelho RA, Lizarraga L, Ponce NMA, et al. Novel gelling pectins from Zea mays husks' agro-industrial residue and their interaction with calcium and iron (II). Bioactive Carbohydrates and Dietary Fibre. 2021;**26**:100273. DOI: 10.1016/J.BCDF.2021.100273

[56] Madhuvanthi S, Selvapriya K, Nirmala RA, et al. Extraction and characterization of pectin derived from underutilized papaya seeds as a value-added product. Journal of Applied and Natural Science. 2022;**14**:127-132. DOI: 10.31018/JANS.V14I1.3269

[57] Kumar M, Potkule J, Tomar M, et al. Jackfruit seed slimy sheath, a novel source of pectin: Studies on antioxidant activity, functional group, and structural morphology. Carbohydrate Polymer Technologies and Applications. 2021;**2**:100054. DOI: 10.1016/J.CARPTA.2021.100054

[58] Chen R, Ratcliffe I, Williams PA, et al. The influence of pH and monovalent ions on the gelation of pectin from the fruit seeds of the creeping fig plant. Food Hydrocolloids. 2021;**111**:106219. DOI: 10.1016/J.FOODHYD.2020.106219

[59] Liu H, Deng L, Dai T, et al. Emulsifying and emulsion stabilization mechanism of pectin from Nicandra physaloides (Linn.) Gaertn seeds: Comparison with apple and citrus pectin. Food Hydrocolloids. 2022;**130**:107674. DOI: 10.1016/J.FOODHYD.2022.107674

[60] Liu HM, He MK, Yao YG, et al. Pectic polysaccharides extracted from sesame seed hull: Physicochemical and functional properties. International Journal of Biological Macromolecules. 2021;**192**:1075-1083. DOI: 10.1016/J.IJBIOMAC.2021.10.077

[61] Selkälä T, Suopajärvi T, Sirviö JA, et al. Surface modification of cured inorganic foams with cationic cellulose nanocrystals and their use as reactive filter Media for Anionic dye Removal. ACS Applied Materials & Interfaces. 2020;**12**:27745-27757. DOI: 10.1021/acsami.0c05927

[62] Duan X, Zhu Y, Shu C, et al. Extraction of pectin from Satsuma mandarin Peel: A comparison of high hydrostatic pressure and conventional extractions in different acids. Molecules. 2022;**27**:3747. DOI: 10.3390/MOLECULES27123747

[63] Kumar M, Tomar M, Saurabh V, et al. Delineating the inherent functional descriptors and biofunctionalities of pectic polysaccharides. Carbohydrate Polymers. 2021;**269**:118319. DOI: 10.1016/J.CARBPOL.2021.118319

[64] Alpizar-Reyes E, Concha JL, Martín-Martínez FJ, et al. Biobased spore microcapsules for asphalt self-healing. ACS Applied Materials & Interfaces. 2022;**14**:31296-31311. DOI: 10.1021/acsami.2c07301

[65] Eivazzadeh-Keihan R, Noruzi EB, Aliabadi HAM, et al. Recent advances on biomedical applications of pectin-containing biomaterials. International Journal of Biological Macromolecules. 2022;**217**:1-18. DOI: 10.1016/J.IJBIOMAC.2022.07.016

[66] Ciriminna R, Fidalgo A, Scurria A, et al. Pectin: New science and forthcoming applications of the most valued hydrocolloid. Food Hydrocolloids. 2022;**127**:107483. DOI: 10.1016/J.FOODHYD.2022.107483

[67] Picot-Allain MCN, Ramasawmy B, Emmambux MN. Extraction, characterisation, and application of pectin from tropical and sub-tropical fruits: A review. Food Reviews International. 2020;**38**:282-312. DOI: 10.1080/87559129.2020.1733008

[68] Khan AA, Butt MS, Randhawa MA, et al. Extraction and characterization of pectin from grapefruit (Duncan cultivar) and its utilization as gelling agent. International Food Research Journal. 2014;**21**:2195-2199

[69] Rodsamran P, Sothornvit R. Microwave heating extraction of pectin from lime peel: Characterization and properties compared with the conventional heating method. Food Chemistry. 2019;**278**:364-372. DOI: 10.1016/J.FOODCHEM.2018.11.067

[70] Dash KK, Ali NA, Das D, et al. Thorough evaluation of sweet potato starch and lemon-waste pectin based-edible films with nano-titania inclusions for food packaging applications. International Journal of Biological Macromolecules. 2019;**139**:449-458. DOI: 10.1016/J.IJBIOMAC.2019.07.193

[71] Almasi H, Azizi S, Amjadi S. Development and characterization of pectin films activated by nanoemulsion and Pickering emulsion stabilized marjoram (Origanum majorana L.) essential oil. Food Hydrocolloids. 2020;**99**:105338. DOI: 10.1016/J.FOODHYD.2019.105338

[72] Xu SY, Liu JP, Huang X, et al. Ultrasonic-microwave assisted extraction, characterization and biological activity of pectin from jackfruit peel. LWT. 2018;**90**:577-582. DOI: 10.1016/J.LWT.2018.01.007

[73] Kaushik P, Priyadarshini E, Rawat K, et al. pH responsive doxorubucin loaded zein nanoparticle crosslinked pectin hydrogel as effective site-specific anticancer substrates. International Journal of Biological Macromolecules. 2020;**152**:1027-1037. DOI: 10.1016/J.IJBIOMAC.2019.10.190

[74] Jannasari N, Fathi M, Moshtaghian SJ, et al. Microencapsulation of vitamin D using gelatin and cress seed mucilage: Production, characterization and in vivo study. International Journal of Biological Macromolecules. 2019;**129**:972-979. DOI: 10.1016/j.ijbiomac.2019.02.096

[75] Demir D, Ceylan S, Göktürk D, et al. Extraction of pectin from albedo of lemon peels for preparation of tissue engineering scaffolds. Polymer Bulletin. 2021;**78**:2211-2226. DOI: 10.1007/S00289-020-03208-1/FIGURES/8

[76] Lapomarda A, De Acutis A, Chiesa I, et al. Pectin-GPTMS-based biomaterial: Toward a sustainable bioprinting of 3D scaffolds for tissue engineering application. Biomacromolecules. 2020;**21**:319-327. DOI: 10.1021/ACS.BIOMAC.9B01332/ASSET/IMAGES/LARGE/BM9B01332_0010.JPEG

[77] Cai T, Xiao P, Yu N, et al. A novel pectin from Akebia trifoliata var. australis fruit peel and its use as a wall-material to coat curcumin-loaded zein nanoparticle. International Journal of Biological Macromolecules. 2020;**152**:40-49. DOI: 10.1016/j.ijbiomac.2020.02.234

[78] Sari M, Tamrin KJ, et al. A novel composite membrane pectin from Cyclea Barbata Miers blend with chitosan for accelerated wound healing. Polymer Testing. 2021;**99**:107207. DOI: 10.1016/J.POLYMERTESTING.2021.107207

[79] Jáuregui KMG, Cabrera JCC, Ceniceros EPS, et al. A New formulated stable papin-pectin aerosol spray for skin wound healing. Biotechnology and Bioprocess Engineering. 2009;**14**:(4):450-456. DOI: 10.1007/S12257-008-0268-0

Chapter 3

Agricultural Pectin Extraction in Iranian Experimental Settings

Roshanak Vakili

Abstract

Pectins are belonged to one important group of polysaccharides extracted from the plant cell walls. Commercial pectins are widely used in the cosmetic, pharmaceutical, and food industries, mainly as texturizing, emulsifying, stabilizing, and gelling agents. Due to rich contents of partially esterified galacturonic acid (GalA) found in agri-food waste, the valorization of recovery process needs to be further developed with economic and environmental benefits. Accordingly, in order to maximize the utilization of these residues, some researchers attempted to extract multiple valuable products from plant waste like pectin from mango peel or simultaneously extracted pectin and polyphenols from pomegranate peels, because the simultaneous extraction seems more efficient due to decreased process time and cost. The characteristics and applications of pectins are strongly influenced by their structures depending on plant species, tissues, and extraction methods. This review aims to review the optimal extraction conditions using new promising methods in order to obtain pectin from Iran's Agro waste and assess physicochemical parameters in recent Iranian experimental study designs, including microwave heating processes and ultrasonic treatment.

Keywords: pectin, Iranian studies, extraction methodology, agricultural waste, secondary metabolites

1. Introduction

Plant pectins are complex polysaccharides with an acceptable content of galacturonic acid (GalA) determined as 65% for commercial purposes like pectin from apple pomace, citrus peels, or sugar beet pulp [1, 2] with different applications based on low (<50%) and high (>50%) methoxyl pectin [3]. High- and low-methylated pectins are often applied according to their different DE, physicochemical properties, and applications; for example, the latter are used in low-calorie products like dietetic jams and jellies [2].

There are various potential sources in pectin production like agricultural wastes [1], for instance, the extracted pectin from mango peel, pomegranate peels, or sour cherry pomace through different extraction procedures and their physicochemical, structural, and functional properties of the extracted compounds have been studied in the literature [3].

Conventionally, pectin can be easily extracted through a cheap and time-consuming acidic hot water extraction (HWE) procedure based on mineral acids like hydrochloric acid, nitric acid, and sulfuric acid [1, 2, 4] based on some factors like raw material, the type of purposed pectin, and manufacturer's instruction [1]. Then, the pectin is recovered by precipitation using ethanol with higher extraction yield. But unfortunately, mineral acids cause serious toxicity and hazardous effect toward the environment, and organic acid like citric acid can be an alternative to this problem [2].

Hence, the extracted pectin is affected by degradation in both quantity and quality aspects with the consequent huge pollutant effluent [1, 4]. It is required to investigate highly efficient and eco-friendly alternative procedures with different plant sources and experimental settings through various optimization protocols for extraction processes [1]. An eco-friendly extraction process can maximize the use of agricultural waste, minimize the volume of the remained waste, and produce the valuable compounds from the global increased agricultural waste in order to have the sustainable waste valorization and the highest utilization of food waste. For example, a large quantity of the industrially processed fruit converts into a large volume of the inevitably produced pomace. The fruit waste is usually discarded, while this valuable source of phenolics and pectin can be extracted in order to increase the financial benefits of production units and reduce the volume of fruit waste and the subsequent environmental problems [5].

These methods include pulsating hydrodynamic action, microwave-assisted extraction (MAE), ultrasound-assisted extraction (UAE), and ultrasound microwave-assisted extraction (UMAE) which lead to increased extraction yield and rate, reduced process time and costs, as well as low environmental hazards aimed to obtain the desirable pectin stability and minimized pectin degradation [1, 4]. Other MAE advantages include reduced wastewater with lower use of organic solvents, increased pectin purity, improved heating rates, decreased equipment sizes, and greater control on the extraction parameters during the process [4]. Based on the cavitation effect of ultrasound waves, eco-friendly UAE methodology is based on the increased destruction of plant cell wall, better penetration, and increased rate of mass transferring with lower consumed solvents and energy [6]. The integrated ultrasonic-microwave synergistic extraction (UMSE) utilizes high energy produced in both devices and removes the defects of HWE, MAE, and UAE at an atmospheric environment with low temperature [4].

In the food industry, serious challenges for the environment are developed with massive quantities of agri-food residues and wastes, including peel, husk, seed, pomace, etc., with the loss of extractable and re-usable valuable compounds at disposal. The designed procedures can prevent the depletion of natural resources by enhancing the economic agricultural opportunities for the rural livelihoods [5].

The average annual agricultural waste is estimated as 35% in Iran with different proposed management procedures. The highest pectin content is found in premature fruits. During ripening, pectin esterase and pectinase enzymes make the pectin percentages decreased, and hence fruits gradually soften. Chemical compounds, structure, and percentage of pectin vary in various plants. Pectin plays a role in food industries as a colloidal additive, thickening/gelling agent, stabilizer, immobilizer, condenser, and emulsifier for traditional use in the production of marmalades, jams, and fruit jellies [7, 8]. According to the experimental literature, industrial pectin can be potentially extracted from the noncommercial sources of agricultural wastes from different fruits and vegetables including peach pomace, watermelon rind, pomegranate peel, papaya peel, sisal waste, pumpkin, soy hull, sunflower oilseed, lemon sour, banana peel, husk of blackberry tree branch, grapefruit peel, Akebia trifoliata husk, peanut, cocoa bean husk, grape pulp, golden kiwi, tomato, carrot, pistachio green husk, and eggplant peel/cap [2, 7].

For example, orange can play an important role in the industry and economy of Iran with producing 2.3 Mt./y in 2019. Citrus peels contain about 20–25% pectin as one of the rich sources of commercial pectin. One of the chemical pretreatments for orange wastes is dilute acid treatment with high pectin recovery in order to hydrolyze it to GalA and other sugars and dissolve the main part of hemicellulose. The valorization of orange waste is oriented on (i) direct utilization (as fertilizer and animal feed), and (ii) extraction of pectin, enzymes, and bioactive compounds with designed biorefinery systems. The relevant studies were oriented toward (a) ultrasounds and microwave treatments to obtain pectin and limonene extraction from citrus waste and (b) to cut the high price of enzyme and its long reaction time in order to eliminate the non-eco-friendly enzymatic hydrolysis [8].

2. Structural and functional properties

Physicochemical structure and functional properties are studied according to (i) gelling properties, (ii) water/oil-holding capacity (WHC/OHC), and (iii) emulsion. The structures are represented by some parameters (molecular weight, MW; degree of esterification, DE; GalA content, and monosaccharide composition) which depend on plant sources and extraction methods and determine the final application. Regarding gelling properties, high-molecular-weight pectins ($\geq$ 300 kDa) can produce a kind of gel which shows network structures with high mechanical strength, rupture strength, and viscosity. In low-molecular-weight pectins, increased medium acidity and higher extraction time are observed. In high-methylated pectin, the gel is formed faster even at acidic pH due to reduced electrostatic repulsion. Low-methylated pectin is able to form a gel by strongly binding divalent ions. Moreover, distribution of non-methoxylated GalA in regular blocks results in gels with a better gelation ability and a stronger mechanical strength. The degree of methoxylation strongly affects Ca-pectin gel properties, and subsequently, acetylation of GalA results in reduced Ca-binding sites and unfavorable gel formation. High active Ca-binding sites induce the percolating network structures and improved rheological properties such as faster formation kinetics, enhanced viscosities, and higher elastic modulus. Regarding WHC/OHC, the hydrophobic/hydrophilic pectin constituents, total charge density and their functions affect texture through the interaction between food product components. High-OHC pectin plays a role in stabilizing or emulsifying in producing high-fat meat foods. WHC shows the hydration ability based on the OH group. The high absorption of water in pectin reduces the syneresis rate in yogurts and dairy desserts [9].

Regarding emulsion, pectin can increase the viscosity of the aqueous phase partly due to homogalacturonan domains with the contribution to emulsion stabilization and higher solution viscosity. The hydrophilic and hydrophobic groups with different amounts and distribution patterns characterize the solubility and rheological properties of pectin-treated liquid food products. In different pectin extraction procedures, different viscosities are created in aqueous solution, affecting emulsion characteristics. Those protein moieties bonded to pectin arabinogalactan and also pectin methyl, acetyl, and ferulic acid ester contents result in high hydrophobicity with an ability to be adsorbed at the oil-water interface. After emulsion formation, the emulsion instability remains to be limited or prevented based on the carbohydrate domain in pectin structure. Moreover, neutral side chains potentially interact with ferulic acid and/or proteins resulted in emulsion stabilization. The commercial pectins produced from citrus peel and apple pomace did not show strong emulsifying properties compared to sugar beet pectin which has the higher protein and ferulic acid contents [9].

3. Iranian literature review

The characteristics and applications of pectins are strongly influenced by their structures depending on plant species and tissues, as well as extraction methods. The aim of this review is therefore to highlight the structures of pectins and the various methods used to extract them, including conventional ones but also microwave heating, ultrasonic treatment, and dielectric barrier discharge techniques, assessing physicochemical parameters which have significant effects on pectin characteristics and applications as techno-functional and bioactive agents [9].

The amount and composition of secondary metabolites produced in plants are controlled significantly by environmental factors, including temperature, carbon dioxide, lighting, ozone, soil water, soil salinity and soil fertility, and climate change which make them to accumulate lower or higher in plants [10, 11]. The adaptation of plant morphology, anatomy, and physiological functions to the changes in biotic and abiotic may influence the accumulation of secondary metabolites. The pathways of secondary metabolites and their regulation are highly susceptible to environmental stresses due to the alteration observed in the involved gene expression [12]. The secondary metabolites play a variety of functions in plant growth and developmental processes, immunity and defense, and finally interaction with environmental stresses [13]. It demonstrates that the multifunctionality of plant secondary metabolites drives interactions between abiotic and biotic factors, with potential consequences for plant resistance in variable environments [14]. The plant has to produce a specified quantity and quality of secondary metabolites to encounter the environmental stress that determines the adaptability and availability of plant in a particular region [11]. In other words, external factors can adversely affect some process associated with biosynthesis of secondary metabolites that ultimately leads to variation in their overall phytochemical profiles, which play important roles in the production of bioactive substances [15].

Therefore, it is essential to perform repeated study designs on plants based on different geographical regions over the world. Here, we have reviewed briefly recent Iran-affiliated studies on pectin extraction procedures performed with agri-food wastes grown in different climatic and geographical settings in order to obtain pectin profiles.

In Kashani et al. [16], three variables were studied in related to their effects on pectin yield, GalA percentage, and DE of pectin including temperature (35, 65, and 95°C), time (40, 120, and 200 min), and pH (1, 2, and 3) with the extracted samples obtained from the potato peels using the acidic or citric acid extraction method in order to optimize the extraction condition profile based on the response surface method. Results of potato peel showed that pectin yield, GalA percentage, and DE ranged 7.15–14.87%, 14.45–36.37%, and 15.35–41.82% in 15 extraction treatments. The physicochemical properties were compared among the potato peel pectin, commercial citrus pectin, and commercial apple pectin according to pectin flow behavior tests at different concentrations, Fourier transform infrared (FT-IR) spectrum, and Mw. In potato peel pectin, the optimized single-independent variables showed that the highest extraction yield was 14.87% with the highest percentage of GalA as 36.37% at 95°C, 120 min, and pH 1.0. Also, the highest DE was 41.820% at 65°C, 40 min, and pH 3.0. Simultaneous optimization for both pectin yield and GalA showed that the highest pectin yield was 15.23% with favorite GalA as 38.0712% at 95°C, 200 min, and pH 1.0. The highest stability of extracted pectin emulsion obtained from potato peel was at 4°C on the first day compared to the stability at 23°C on the 30th day. According to FT-IR results, the strong absorption seen between 3200 and 3500 cm^{-1} was related to the intracellular/extracellular vibration of the hydrogen bonds in the GalA polymer. At increased pectin concentration (0.1–2%), the viscosity was increased and Newtonian behavior was observed in all samples with flow index of 1. In potato peel, Mw of the extracted pectin was 53.46 kDa after 30 days of storage under optimal conditions at 4 and 23°C with emulsion stability (ES) of 85.1 and 63.1, respectively. Therefore, the produced pectin obtained optimally from agricultural wastes using citric acid procedure can be introduced to the market with Newtonian behavior and optimal gel grade.

In Kazemi et al. [17], the cantaloupe rind was effectively valorized into food-grade pectin by an environmentally friendly MAE process without the application of mineral acid. Then, the extraction factors were optimized by Box-Behnken design (BBD), and the extracted pectin was characterized according to various physicochemical, structural, functional, and bioactivity properties. Four variables of the extraction process were successfully optimized (microwave power: 700 W, irradiation time: 112 s, pH 1.50, and liquid-solid ratio (LSR): 30 mL/g) with a yield of 181.4 g kg^{-1}. After analysis, it was found that the isolated pectin was a high-methylated GalA-rich sample (703.4 g kg^{-1}) with an average Mw of 390.475 kDa. Also, the isolated pectin was a high-potential sample with favorite functionality and antioxidant ability in comparison with commercial citrus pectin according to FT-IR, Hydrogen-1 nuclear magnetic resonance (1H-NMR), and X-ray diffraction (XRD) spectroscopies. The main functional groups, structural characteristics, and crystallinity showed that the assayed samples had a significantly higher value of OHC, emulsifying capacity (EC), ES, α-diphenyl-β-picrylhydrazyl (DPPH), and 2,2′-azino-bis(3-ethylbenzothiazoline-6-sulfonic acid) (ABTS) scavenging activity, and reducing power assay with very promising quantity (yield) and quality values. The potential of MAE process included the remarkable reduction of both production time (instead of hours into minutes) and energy consumption, economically and environmentally promoted productivity. In industrial scale, it is necessary to evaluate the MAE process to demonstrate much better pectin yield, operating costs, and environmental burdens. Moreover, the repeatability and the functionality of isolated pectin are essential in future studies.

In Peighambardoust et al. [18], the comparative study objective was to identify the properties of beet pectin based on a novel and ecofriendly technology of subcritical water extraction (SWE) instead of traditional procedures in order to maximize pectin extraction efficiency. The advanced modeling procedures were applied including response surface methodology and optimization of operational parameters. The results indicated a promising scalable approach for converting the beet waste to pectin based on SWE with improved pectin properties. The optimal conditions for obtaining the highest pectin yield were determined using the central composite design for both comparative methods. In traditional procedure, the temperature, time, pH, and pectin recovery yield were 90°C, 240 min, 1, and 20.8%, respectively. In the subcritical water extraction, LSR was 30% (v/w) at temperature of 130°C for 20 min with a comparable yield of 20.7%. The effect of obtained pectin samples on viscoamylograph pasting and differential scanning calorimetry (DSC) thermal parameters of corn starch was assessed. The GalA content, degree of methylation, acetylation, and ferulic acid content were higher in the pectin obtained using SWE, while their Mw was lower. Both pectin samples have similar chemical groups according to FT-IR with similar colors. At lower concentrations (0.5–1%), pectin solution obtained in both techniques nearly indicated a Newtonian behavior according to the rheological measurements. The addition of both pectin samples to corn starch decreased both pasting (except hot paste viscosity) and DSC gelatinization (peak temperature or Tp, conclusion temperature or Tc, and Tc – To (onset temperature) of starch parameters, while increased ΔH to higher values. The To was minimally affected after the addition of pectin. Brabender viscoamylograph results were in good agreement with the DSC results. SWE was more efficient with the same extraction yield in a much shorter time due to nearly being 12 times faster; therefore, it can reduce pectin extraction time on an industrial scale and also facilitates the achievement of the pectin with improved specifications.

In Vaez et al. [8], a multipurpose platform was developed to dilute acid pretreatment as a multipurpose process to recover pectin, hydrolyze hemicellulose, and open up the cellulose structure. Some studies used separated fractions of orange wastes divided into pulp and peel, but it is not possible to practically separate them at the industrial scale and according to a conventional laboratory setting it is required to study their potential separately. Vaez et al. [8] designed a dilute acid treatment method on orange waste to extract pectin and fermentable sugars as well as breaking down the recalcitrant structure of the remained lignocellulose. The fermentable sugars dissolved in the supernatant were used for ethanol production without any further procedures for enzymatic hydrolysis. After acid treatment, the pretreated remained solid fraction was used for biogas production. One advantage of their designed biorefinery platform is related to the removal of enzymatic hydrolysis, that is a necessary step in a conventional ethanol production process. Therefore, biogas-ethanol-pectin integrated production is practiced in their study based on a dilute acid treatment procedure on orange waste with sulfuric acid (1% w/v) (94, 100, 140, and 180°C; 60, 30, and 0 min). Finally, the pectin was extracted from the hydrolysate, the liquor was used to produce ethanol, and the pretreated solid was anaerobically digested to produce biogas. The highest pectin extraction yield was 24.7% (w/w) and 23.7% (w/w) from orange peel and pulp fractions, respectively, from the supernatants of liquor treatment at 94°C for 60 min. FT-IR results confirmed the similar characteristics of the extracted pectin to the commercial sample. The GalA content (as pectin purity) was 70.2 and 69.9% from orange peel and pulp, respectively, at the optimal conditions. The acid treatment at 94°C for 60 min achieved a pectin product with approximately 69% of DE compared to approximately 45% in the treatment procedure at 140°C for 30 min. The highest ethanol yields of 81.5 and 82.9% were achieved from orange peel and pulp, respectively, after the acid treatment at 140°C for 30 min. The highest methane yields were 176.8 and 191.8 mL/g as volatile solids (VS) from the untreated orange peel and pulp, respectively. The highest total product value was 2472.9 USD/t orange wastes with dilute acid treatment at 94°C for 60 min. At the optimal conditions related to high pectin production with no enzyme, 244 kg of pectin, 26.5 L of ethanol, and 36 m^3 of methane were obtained from 1 t of orange wastes. If biogas is intended, the treatment procedure of citrus waste is not required. The proposed biorefinery platform can increase the total products value up to 75 times compared to the traditional anaerobic digestion of citrus waste [8].

In Ezzati et al. [19], the extracted pectin of sunflower by-product was obtained using UAE technique. The UAE variables were successfully optimized using BBD optimization process (irradiation time: 30 min, temperature: 33°C, ultrasound power: 400 W) with 11.15% of pectin yield. It was found that the extracted pectin was proved to be a high-purity sample and rich in low-esterified GalA content (72.94%) with long-side galactan branches, arabinogalactan, and arabinan, with an average Mw of 175.353 kDa. The functional groups and structural characteristics were determined by FT-IR, 1H-NMR, and XRD spectroscopies. According to the results obtained from DSC and thermogravimetric analysis (TGA) procedures, the thermal analysis suggested a suitable thermal stability for the extracted pectin. Other functional parameters were measured including the solubility, WHC, OHC, EC, ES (in different conditions), foam capacity, foam stability, DPPH, and ABTS inhibitions for assaying antioxidant properties and reducing power assay in order to prove the higher value of the extracted pectin for the potential of replacing to commercial food ingredients. Therefore, the obtained pectin can be used as a high-quality pectin sample with good functional and technological properties in pharmaceutical or food industries.

Hosseini et al. [3] studied the optimized MAE conditions for the simultaneous recovery of pectin and phenolic compounds from sour cherry pomace. An annual production of >109,000 tons in Iran makes this country as the seventh producer of sour cherry in the world, and the residual pomace that is discarded as a side product in food industry and is rich in different polysaccharides such as pectin can lead to environmental problems; but the reuse of fruit wastes for their production has high benefits for nutritional and environmentally friendly. They found that the highest yield of pectin (14.65 ± 0.39%, Mw: 472.977 kDa) was obtained with microwave power of 800 W, irradiation time of 300 s, pH 1.0, and LSR of 20 v/w. The structural analysis indicated that the obtained supernatant was rich in high-methoxyl pectin with amorphous structure. The moisture, ash, and protein contents as well as total carbohydrates were nearly 8.32, 3.73, 1.41, and 26.43%, respectively. The high purity was proved from approximate GalA content of 72.86%, suitable thermal stability was obtained due to degradation temperature of 252.15°C, and also high-methoxyl pectin was proved from DE of 68.37 ± 2.78%. According to the FT-IR, 1H-NMR, and XRD analysis, the obtained sample was rich in esterified polygalacturonic acid with an amorphous structure.

In Kazemi et al. [20], BBD was applied to optimize the conditions of the UAE and heating extraction (HE) procedures according to pectin yield from pistachio green hull as a response. At optimal condition, the pectin extraction yield of the UAE (ultrasound power: 150 W, pH 1.5, time: 24 min) and HE (temperature: 70°C, time: 90 min, LSR: 40 v/w) methods were 12.0 ± 0.53% and 10.3 ± 0.75%, respectively. Also, the GalA content of pectin was about 59.33 and 75.11% in UAE and HE methods, respectively, which was higher in the latter. Both methods were able to achieve good emulsifying activity, WHC and OHC. Moreover, total phenol content (TPC) and antiradical activity (DPPH radical scavenging) of pectin samples were higher in HE method compared to UAE procedure. But the surface tension value was lower in the former which was in agreement with the results obtained from foaming properties. The decreased surface tension with the increased TPC can be probably attributed to the accumulation of phenolic compounds in the air-water interface with a resulted increase in the surface pressure and finally a resulted decrease in surface tension. In the HE method, DE was higher compared to the pectin produced in the UAE method according to FT-IR and 1H-NMR analysis. The results of XRD patterns revealed an amorphous structure with some crystalline portions in the fruit pectin. The surface morphology showed more surface roughness for pectin obtained from the HE method compared to the UAE method. According to the rheological properties of pectin solutions, G′ and G″ of pectin extracted from the HE method in 2% w/v were much higher than the UAE method. Both samples showed similar elastic behavior in high frequencies, while pectin sample obtained from HE method had a viscous behavior in low frequencies. The UAE method had significantly increased pectin yield with a lower processing time and consumed solvent volume; however, pectin sample from HE method presented a better quality. Therefore, there are some limitations and disadvantages for running UAE procedure in industrial scale.

Khodaiyan and Parastouei [5] investigated pectin extraction from black mulberry pomace based on an eco-friendly extraction process of MAE procedure. The variables were successfully co-optimized using BBD, and then the optimized condition yielded about 10.95% pectin as response. The compounds produced under optimum conditions (microwave power: 700 W, irradiation time: 300 s, pH 1.42, and LSR: 20 mL/g) were characterized based on physicochemical, structural, and functional properties. The increased production of global agricultural waste urges critical attention to the concept of sustainable waste valorization and maximum utilization of food waste; therefore, the production of several products from waste has attracted high interests. According to the physicochemical analysis, there was a highly esterified amorphous pectin (DE: 62.21%) with an average MW of 620.489 kDa based on XRD analysis and a highly esterified GalA content of 70.15% with further confirmation based on FT-IR and 1H-NMR spectroscopies. Also, the DSC showed higher thermal stability for the assayed pectin than commercial pectin (degradation temperature: 251.82°C). The designed procedure provides a promising management of black mulberry waste generated in food industry with high quality for being applied as natural ingredients in various food and pharmaceutical products. The final aim is to maximize the waste use for the production of the valuable compounds and minimize the waste volume.

Nouri and Mokhtarian [7] studied on the pectin processed from walnut green husks and found them as good sources for pectin extraction. According to the response surface statistical methodology, they assessed extraction efficiency rate, DE, and GalA of the obtained pectin in different pH (1.0, 1.5, 2.0), temperature (60, 70, 80°C), and process time (60, 90, 120 min) values. The optimal samples were selected, and total ash, MW, emulsifier, rheological, and FT-IR spectroscopy assessments were performed. The highest efficiency rate (25.84%) was obtained at optimal conditions (pH 1.62, 80°C, and 120 min). The highest DE (63.19%) occurred at optimal conditions (pH 2.0, 72.92°C, and 87.27 min, range: 52.30–60.20). The GalA proportion indicating purity of pectin was normal. The highest GalA (68.53%) was recorded at pH 1.44 and 72.92°C in 93.33 min. Some viscous and pseudoplastic behaviors were assayed with the extracted pectin. According to FT-IR spectral diagrams, the optimal pectin samples have shown the presence of GalA as a rich source of pectin.

Gharibzahedi et al. [4] studied on the comparative pectin extraction procedures from common fig skin, including HWE, UAE, MAE, and UMAE. The results showed that UMAE (11.71%) significantly obtained a more extraction yield than MAE (9.26%), UAE (8.74%), and HWE (6.05%). The UMAE-pectin with the highest GalA content (76.85%) and MW (6.91 × 10 3 kDa) had the highest emulsifying activity (61.2–61.3%) and ES (94.3–95.2%) with a monomodal droplet size distribution at both cold and ambient storage temperatures. A non-Newtonian shear-thinning behavior was recorded at 1.5–3.0% pectic solutions. XRD analysis showed noncrystalline pectin extracted by UMAE. FT-IR spectroscopy and high-performance liquid chromatography (HPLC) photodiode array detector proved that both conventional and novel extraction technologies do not change the chemical structure and monosaccharide composition of pectin significantly. The UMAE at operating conditions (pH 1.4, 1:20 g/mL CFS[1]/water, sonication time: 25 min, irradiation time: 3.5 min and microwave power: 600 W) was proved to be a successful strategy to extract high-MW pectin from fig skins with the highest extraction yield, total GalA, viscosity of pectic solution, emulsifying activity, and ES at different assay conditions. The pectin functionality for food-grade emulsions was also proved because oil-in-water emulsions stabilized

[1] cubic feet per second.

with fig skin pectin extracted by UMAE had the lowest droplet size with a monomodal size distribution. All extracted pectin samples had a DE being lower than 50 and can be applied in stable formulations of many low-sugar dietary foods. A pseudoplastic flow behavior was observed at the high concentration of pectin. The main functional groups and monosaccharides determined based on FT-IR spectroscopy and pulsed amperometric detection (HPLC-PAD), respectively, are clues for the extracted polygalacturonic acid-rich pectin. However, it is essential to conduct an optimization study in order to find the functional conditions for finding the best UMAE extraction method and obtaining the highest extraction yield of fig skin pectin.

Hosseini et al. [6] performed a study for optimization and characterization of pectin extracted from sour orange peel by UAE procedure. Their aims were as follows:

i. new UAE-assisted optimization of fruit pectin in order to find the effects of various extraction factors on pectin yield and properties,

ii. structure, monosaccharide, and chemical compositions of the extracted pectin in related to extraction methodology,

iii. physicochemical and functional properties of fruit pectin.

In this work, BBD was applied with three variables (ultrasound power, irradiation time, and pH) for pectin extraction optimization from sour orange peel in three levels by ultrasound waves. The physicochemical, structural, and functional properties of fruit pectin were evaluated in optimal extraction point. According to the obtained results, the highest extraction yield was 28.07 ± 0.67% in optimal conditions (ultrasound power: 150 W, irradiation time: 10 min, pH 1.5). Also, ash, moisture, and protein contents of fruit pectin were 1.89 ± 0.51%, 8.81 ± 0.68%, and 1.45 ± 0.23%, respectively. Moreover, 65.3% of the extracted pectin was GalA with approximately 72% of total neutral sugars as galactose according to HPLC findings which showed the fruit pectin has a suitable purity. In the optimized pectin, there are TPC of 39.95 ± 3.13 mg gallic acid equivalents/g pectin, the surface tension of 46.56 ± 0.23 and 42.14 ± 0.61 mN/m in concentrations of 0.1 and 0.5%w/v, and WHC and OHC of 3.10 ± 0.12 and 1.32 ± 0.21 g water or oil/g pectin, respectively. In addition, the emulsifying activity of fruit pectin extracted by ultrasound waves was higher than those samples from other sources, and emulsions were more stable in low temperature. Moreover, DE of 6.77 ± 0.43% was proved the fruit pectin to be a low-methoxyl pectin according to FT-IR and 1H-NMR analysis. Therefore, the procedure with ultrasound waves showed a high efficiency based on quantity/quality of the extracted pectin. These waves improve the destruction of plant cell wall by their cavitation effect and increase the rate of mass transferring resulted in the higher extraction yield of pectin in a shorter extraction time [6].

In Kazemi et al. [21], eggplant peel was used for pectin extraction through UAE technique. The optimization process was carried out using BBD in order to optimize the extraction process factors, and the results showed that the highest experimental extraction yield (33.64 ± 1.12 g/ 100 g, the predicted yield: 35.36 g/100 g) was achieved with optimal conditions (ultrasound power: 50 W, irradiation time: 30 min, pH 1.5). The assay of chemical, physicochemical, functional, and structural pectin features indicated that it was rich in GalA (66.08 g/100 g) and has both high DE (61.22%) and TPC (96.81 mg GAE/g pectin) and both low ash and protein contents. Also, the extracted pectin showed favorite measurements of functional properties including WHC, OHC, emulsifying and foaming properties, and antioxidant activity.

In addition, FT-IR and 1H-NMR spectroscopy proved a high-methylated pectin structure in the obtained samples. Moreover, assaying DE suggested that the extracted pectin is in the group of high-methoxy pectin with further confirmation based on those measurements from FT-IR and 1H-NMR spectroscopy. XRD pattern proved a high crystallinity for eggplant pectin. Given the high extraction yield and favorite properties, the eggplant peel pectin can be a good replacement for commercial pectin.

In Jafari et al. [2], the central composite design was used with four variables in five levels to determine the effects of pH (0.5–2.5), temperature (50–90°C), heating time (30–150 min), and LSR (10–50 v/w) on both yield and DE of the extracted carrot pectin. The highest extraction yield of pectin was 15.6 ± 0.5% at optimal conditions (90°C, 79.8 min, LSR of 23.3 v/w, pH 1.3) which was close to the predicted values (16.0%). According to the obtained findings, the extracted pectin was proved to be a low-methoxylated pectin (DE: 22.1–51.8%) with a favorite emulsifying activity (60.3%), viscosity at a wide range of frequencies (0.1–50 Hz, 1% w/v), and pseudoplastic flow behavior at the same concentration. With the optimal extraction conditions, the GalA content and emulsifying activity were 75.5 and 60.3%, respectively; moreover, the emulsions had a high stability (80.4–80.3%, 74.7–74.4%) at two different storage temperatures (4 and 23°C) after 1 and 30 days, respectively.

Bagherian et al. [1] performed a comparative study on the conventional and microwave- and ultrasound-assisted methods for the extraction of pectin from grapefruit. In this study, the effect of microwave power and heating time was assayed on both pectin yield and quality in grapefruit. It was found that the highest pectin yield was 27.81% (w/w) at 6 min and 900 W. It was observed that pectin yield, the GalA content, and DE increased with the increased microwave power and heating time. But Mw decreased with an increase in heating time; however, the effects of power on Mw were dramatically more than heating time. In addition, laboratory studies on the extraction of pectin treated with high-intensity ultrasound were carried out. The effects of temperature and time on quality and quantity of extracted pectin were investigated. The highest yield was for sonication time of 25 min (17.92%) in a constant bath temperature of 70°C. Furthermore, before applying MAE the grapefruit solution was treated by a preliminary ultrasonic heating and a higher yield was proved. Intermittent sonication was so efficient than continuous procedure. The studied parameter in microwave extraction included microwave field power and heating time for improving both qualitative and quantitative characteristics of extracted pectin. Finally, 2 min of microwave heating was able to induce the same amount of pectin obtained as with 90 min of conventional extraction procedures. On the other hand, sonication was performed with water bath, and the effect of sonication time and bath temperature were measured on the pectin extraction in order to find the optimal factors. The conventional procedure was not able to compete with sonication method being 3 times faster. Also, when the ultrasound pretreatment was performed before microwave heating, better results were obtained than MAE.

4. Methodology

4.1 Case study No. 1

An optimal biorefinery development for pectin and biofuels production from orange wastes without enzyme consumption—A case study by Vaez et al. [8].

4.1.1 General procedure

The orange wastes were treated with sulfuric acid (1% w/v) at 94, 100, 140, and 180°C for 60, 30, and 0 min in order to extract pectin from the hydrolysate. The highest yield of pectin extraction was 24.7% (w/w) and 23.7% (w/w) in orange peel and pulp, respectively, in the supernatants at 94°C for 60 min. The results of FT-IR confirmed that the characteristics of the extracted pectin were similar to the commercial product. Regarding the pectin purity, the GalA content was 70.2 and 69.9% in orange peel and pulp, respectively. The acid treatment (94°C, 60 min; 140°C, 30 min) showed a DE being higher than 69% and less than 45%, respectively. At the optimal conditions, high production of pectin was 244 kg/t of orange wastes without enzyme procedure.

4.1.2 Preparation of material

Fresh Thomson navel orange (*Citrus sinensis*) from northern Iran was used in this study. After the fruits were washed, they were juiced, and the separated pulp and peel were dried at ambient temperature for 2 weeks and then they were grinded. Those particles being sized as 60.85–0.180 mm (20–80 mesh) were collected and kept in sealed plastic bags for further analysis at room temperature. The prepared orange peel and pulp contained 92.6 and 93.6% total solids (TS) and 87.3 and 89.9% volatile solids (VS), respectively.

4.1.3 Dilute acid treatment

Sulfuric acid solution (1% w/v) was applied with different temperatures (100, 140, and 180°C) and times (0, 30, and 60 min) in 101SSHPR[2] reactor warmed up in an oil bath. In order to extract pectin, the optimal acid treatment condition was encountered at 94°C for 60 min with 140 mL of sulfuric acid (1% w/v) and 10 g of loaded orange peel/pulp in the reactor. Then, the temperature was quickly decreased using an ice bath. The suspension was filtered for the solid residues that rinsed with distilled water to remove chemicals. Then, they were freeze-dried and kept at room temperature, but the liquid fraction was kept at −20°C for pectin extraction and purification.

4.1.4 Pectin extraction and purification

For the highest pectin yield, pH was adjusted to 3.5 with sodium hydroxide. The solution was dispersed in ethanol, and the precipitates were obtained after it remained for a night at 4°C with centrifugation (4000 rpm, 20 min). Ethanol at a 70 and 96% (v/v) alcohol content was used for rinsing the separated solids. The mixture was centrifuged (4000 rpm, 30 min) to separate the pectin. The separated pectin was dissolved in deionized water, freeze-dried, and maintained at room temperature. The yield of pectin extraction was calculated using an equation (Bi and P as initial quantity in gram of substrate and pectin, respectively):

$$Y_{pec}(\%) = \frac{P}{B_i} \times 100 \quad (1)$$

[2] 101 stainless steel high pressure reactors.

4.1.5 Analytical methods

4.1.5.1 Substrate characterization

The TS, VS, and extractive contents of the orange wastes were measured according to the National Renewable Energy Laboratory (NREL) protocol. The morphology of the freeze-dried substrates was assayed using gold-coated procedure and scanning electron microscopy (15 kV) in order to observe the acid treatment effect.

4.1.5.2 Pectin characterization

Pectin characteristics (chemical structure, GalA content, and DE) were determined. The chemical structure was assessed by a Fourier transfer infrared spectrometer with a deuterated-triglycine sulfate detector in comparison with commercial pectin (resolution of 1 cm^{-1}, 32 scans in 4400–400 cm^{-1}).

The GalA content was quantified based on Ramos-Aguilar et al. [22]. Pectin (5 mg) was added to 2 mL of concentrated sulfuric acid (98%) and 1 mL of deionized water and adjusted to 10 mL. After an ice bath procedure (10 min), the centrifugation was performed at room temperature (2000 × g, 10 min). The liquid fraction (400 μL) was mixed with 2.4 mL of sodium tetraborate (75 mM in concentrated sulfuric acid) and 40 μL of 4 M potassium sulfamate solution (pH 1.6). The temperature of tubes decreased using indirect contact in an ice bath after insertion in boiling water for 20 min. The M-hydroxy di-phenyl solution in NaOH was added. The absorbance of samples was assayed at 525 nm using a UV-vis spectrophotometry.

The DE was obtained according to Santos et al. [23] as follows: the dry mass of pectin (0.1 g) was dissolved with 3 mL of ethanol 96% in 20 mL of distilled water at 40°C and 100 mL Erlenmeyer flasks were magnetically stirred up. The titration was performed with sodium hydroxide solution (V1 mL for the first neutralization step, 0.1 M) and phenolphthalein indicator for appearing a pale pink color. The sodium hydroxide solution (10 mL, 0.1 M) was added to this neutralized solution and stirred. The hydrochloric acid solution (0.1 M, 10 mL) was added and stirred in order to disappear the pink color completely. The sodium hydroxide solution (V2 mL for the second neutralization step, 0.1 M) was added to neutralize the excess of acid until the solution color changed to pink (V2). The DE of the pectin was determined using V1 and V2 as follows:

$$DE(\%) = \frac{V_2}{V_1 + V_2} \times 100 \quad (2)$$

4.1.5.3 Statistical analysis

Analysis of variance (ANOVA) was applied using the least significant difference (LSD) and Tukey's methods in SAS 9.1.3 software ($p < 0.05$).

4.2 Case study No. 2

High-quality pectin from cantaloupe waste: Eco-friendly extraction process, optimization, characterization, and bioactivity measurements—A case study by Kazemi et al. [17].

4.2.1 Extraction and optimization of pectin

Fresh cantaloupe fruits were washed, peeled, cut, and placed in an oven (60°C, 48 h). The dried pieces were powdered and stored at 25°C. The MAE process was used to produce pectin from cantaloupe powder in a microwave oven according to Kazemi et al. [17]. pH levels were adjusted using organic citric acid. The response was expressed as the extraction yield (g kg^{-1}). For optimization of the MAE process, MAE variables and their levels were selected for pectin extraction from cantaloupe rinds using BBD according to the literature. The effects of MAE-independent variables were optimized and investigated including microwave power (300–700 W), irradiation time (60–180 s), pH (1.5–3.0), and LSR (20–30 mL/g). In order to reduce systematic errors, all experiments were randomly performed, and then the results were put in a polynomial equation in order to predict the optimized condition for the MAE process.

4.2.2 Pectin characterization

Pectin characteristics (physicochemical, structure, functional, and antioxidant assay) were determined. Three experiments were performed for each measurement, and the results were calculated and reported as mean value ± SD[3].

4.2.3 Physicochemical analysis

For moisture content, 1 g of fruit pectin was dried at 105°C for 24 h. Ash content was measured after dry ignition of 1 g of pectin samples (550°C, 6 h) in a muffle furnace. Protein content was determined using the Kjeldahl method (crude protein = N × 6.25). Total carbohydrate content was measured according to the phenol sulfuric acid method. The GalA content was determined using the meta-hydroxydiphenyl method. The DE was measured according to Kazemi et al. [17]. TPC was found on an SP-UV500DB model UV-vis spectrophotometer according to the Folin-Ciocalteu method using a standard curve of gallic acid. TPC was expressed as mg gallic acid equivalent per g of pectin (mg GAE/g pectin). All physicochemical measurements were compared based on the commercial citrus pectin.

Average MW was determined by injection of pectin solution (20–50 μL, 1 mg/mL) high-performance gel permeation chromatography (HPGPC) using an Ultrahydrogel™ column and a RID[4]-10A refractive index detector. For elution, NaNO3 (0.1 M) was used as mobile phase at a flow rate of 1 mL/min and temperature of 35°C. The calculation was based on the calibration curves of dextran standards.

4.2.4 Structural analysis

The isolated pectin was structurally characterized using FT-IR, 1H-NMR, and XRD spectroscopies in comparison with commercial citrus pectin. FT-IR spectrum of the KBr-dispersed samples was recorded on a Bruker Tensor 27 spectrometer by 10 scans at a resolution of 4 cm^{-1} over 4000–600 cm^{-1}. 1 H-NMR spectrum was collected on a 500 MHz Varian Unity Inova spectrometer by eight scans (24°C, 4.0 s). XRD pattern was recorded on a Philips diffractometer (10–80° as 2θ).

---

[3] Standard deviation.

[4] Refractive Index Detector.

4.2.5 Functional characteristics

The functional properties were measured using four methods including WHC, OHC, EC, and ES in comparison with commercial citrus pectin. WHC and OHC were measured according to Kazemi et al. [17] as the mass (g) of water or oil retained by 1.0 g of pectin sample (g/g). EC (24°C) and ES (1 and 30 days, 4°C and 24°C) of emulsions were measured according to Yapo et al. [24].

4.2.6 Antioxidant assay

The antioxidant capacity of pectin solutions was determined in concentrations of 0.1–5 mg/mL using three methods, including DPPH scavenging activity, ABTS scavenging activity, and reducing power assay in comparison with ascorbic acid, beta hydroxy acid (BHA), and commercial citrus pectin at similar concentrations. Also, the half maximal inhibitory concentration values of DPPH and ABTS scavenging activity for each sample were calculated and compared (the required concentration of antioxidant for scavenging the 50% of initial concentration of free radicals).

4.2.7 Statistical analysis

Response surface BBD was used to assess the effects of MAE process variables individually and interactively in terms of pectin yield (56–150 g kg^{-1}) and the highest pectin yield. The experimental data was applied in a second-order polynomial model developed using multiple regression analysis. Also, in order to show a good agreement between the experimental and predicted data, Pareto ANOVA was used. According to the obtained results, a fitted developed model could explain the suitable relationship between process variables and response.

5. Conclusions

In food industry, the functional properties of pectin are influenced by the source, methods, and conditions of extraction (time, pH, LSR, temperature, wave power and frequency, enzymes, and the integrated conditions). Novel extraction methods are focused in studies in order to reduce extraction time/solvent consumption and increase process efficiency/pectin yield. Agri-food wastes can be processed to produce valuable by-products like marketable pectin according to commercial standards. Their significant WHC/OHC and promising emulsifying properties make them as textural ingredients and emulsifiers in food products and pharmaceutical supplements [9]. The green chemistry extraction process for the valorization of industrial food processing waste should be designed based on circular economy concepts in order to increase its benefits [5]. However, further research is needed to understand the synergistic effect of multiple extraction factors using new techniques for the improvement of the productivity of industrial pectin processes from Iran's Agro waste (**Tables 1** and **2**).

Studies published in Persian:

Akbari-Adergani B, Zivari Shayesteh P, Pourahmad R. Evaluation of some functional properties of extracted pectin from pomegranate peel by microwave method. *Journal of Food Technology and Nutrition*. 2021;18:5–16 [In Persian].

Amanpour M, Asefi N. Effect of ultrasound time and acid type on the qualitative properties of extracted high methoxyl pectin from quince fruit. *Iranian Journal of Nutrition Sciences and Food Technology*. 2020;15:71–82 [In Persian].

Azadbakht M, Tabaey MH, Sabet Ahd Jahromi A. The comparison of various methods for extraction and isolation of pectin from *Citrus Decumana* Murry. *Journal of Tabriz University of Medical Sciences (Pharmaceutical Sciences)*. 2003;1:21–28 [In Persian].

Bahramipour M, Akbari-adergani B. Optimization of microwave-assisted extraction of pectin from peaspod by response surface method. *Journal of Food Science and Technology*. 2018;80:349–360 [In Persian].

Darvishi M, Yazdanpanah S. Evaluation and comparison of emulsion, rheological and spectroscopic properties of FT-IR extracted pectin from peel and cap of pumpkin. *Journal of Food Science and Technology*. 2020;17:149–162 [In Persian].

Ebrahimzadeh MA, Azadbakht M. (2006). Extraction of pectin and comparison of yield and degree of esterification and percentage of galacturonic acid in the skin of some citrus. *Journal of Mazandaran University of Medical Sciences*. 2006;16:52–59 [In Persian].

Fathi B, Maghsoudlou Y, Ghorbani M, Khomeiri M. Effect of pH, temperature and time of acidic extraction on the yield and characterization of pectin obtained from pumpkin waste. *Journal of Food Research (AGRICULTURAL SIENCE)*. 2012;22:465–475 [In Persian].

Haji Esmaelli F. Extraction of pectin from sugar beet pulp [master's thesis]. Tehran (Iran): Science and Research Branch. Islamic Azad University; 2002 [In Persian].

Hosseini SS, Khodaiyan F, Barazande S. Extraction and comparison of the physicochemical properties of pectin extracted from pineapple, samsuri and galia melon peels assisted by microwave. *Iranian Journal of Food Science and Nutrition*. 2017;11:71–80 [In Persian].

Jannat B, Oveisi MR, Sadeghi N, Behzad M, Behfar AAA, Hajimahmoodi M, et al. Determination of pectin in sunflower and its application in food industry. *Journal of Food Technology and Nutrition*. 2016;13:25–34 [In Persian].

Karimifard SM. Separation of valuable chemicals from citrus waste and conversion of the produced wastes to biogas [master's thesis]. Isfahan (Iran): Isfahan University of Technology; 2013 [In Persian].

Keramat J, Kabir GM, Ghenaati B. Qualitative and quantitative study of pectins extracted from the orange juice concentrate production process waste. *Agricultural Science and Technology and Natural Resources*. 2002;6:141–148 [In Persian].

Mesbahi GR, Jamalian J. Pectin extraction from sugar beet pulp and its application in food products. *Journal of Agricultural Science and Technology of Agriculture and Natural Resources, Water and Soil Science*. 2002;6:125–138. [In Persian].

Mirmajidi A, Babaei B. Extraction of pectin by pre-hydrolysis method from some commercial type of sugar beet. *Research in Agricultural Engineering*. 2007;9:103–114. [In Persian].

Mosayebi V, Emam-Djomeh Z, Tabatabaei Yazdi F. Optimization of extraction conditions of pectin by conventional method from black mulberry pomace. *Quarterly Journal of Food Science and Technology*. 2017;62:341–356 [In Persian].

Nateghi L. Optimization of extraction conditions of pectin from *Aloe Vera* leaf. *Food Technology and Nutrition*. 2021;18:95–115. [In Persian].

Nateghi L, Ansari S, Shahab Lavasani AR. Investigation of yield and physicochemical properties of pectin extracted from eggplant peel. *Food Science and Technology*. 2017;73:13–30 [In Persian].

Nateghi L, Ansari S. Investigation of yield and physicochemical properties of pectin extracted from eggplant cap. *Journal of Modern Food Technologies*. 2017;5:219–239 [In Persian].

Saberian H. Optimization of pectin extraction from orange waste by ohmic and enzymatic novel methods [Ph.D Thesis]. Tehrn (Iran): Tarbiat Modares University; 2017 [In Persian].

Zarei M. Effect of extraction temperature on qualitative characteristics extracted pectin from orange peel. 18th National Congress on Food Technology, Mashad, Iran; 2008 [In Persian].

Studies published in English:
Akbarpour V, Hemmati K, Sharifani M. Physical and chemical properties of pomegranate (*Punica granatum* L.) fruit in maturation stage. *American-Eurasian Journal of Agriculture and Environmental Sciences*. 2009;6:411–416.
Asgari K, Labbafi M, Khodaiyan F, Kazemi M, Hosseini SS. High-methylated pectin from walnut processing wastes as a potential resource: ultrasound assisted extraction and physicochemical, structural and functional analysis. *International Journal of Biological Macromolecules*. 2019;152:1274–1282.
Chaharbaghi E, Khodaiyan F, Hosseini SS. Optimization of pectin extraction from pistachio green hull as a new source. *Carbohydrate Polymers*. 2017;173:107–113.
Faravash RS, Ashtiani FZ. The influence of acid volume, ethanol to extract ratio and acid washing time on the yield of pectic substances extraction from peach pomace. *Food Hydrocolloids*. 2008;22:196–202.
Gavlighi HA, Tabarsa M, You S, Surayot U, Ghaderi-Ghahfarokhi M. Extraction, characterization and immunomodulatory property of pectic polysaccharide from pomegranate peels: Enzymatic vs. conventional approach. *International Journal of Biological Macromolecules*. 2018;116:698–706.
Gharibzahedi SMT, Smith B, Guo Y. Ultrasound-microwave assisted extraction of pectin from fig (*Ficus carica* L.) skin: Optimization, characterization and bioactivity. *Carbohydrate Polymers*. 2019;222:114992.
Hashemi A, Razzaghzadeh S. Investigation on the possibility of ensiling cucurbit (*cucurbita pepo*) residues and determination of best silage formula. *Journal of Animal and Veterinary Advances*. 2007;6:1450–1452.
Hosseini AS, Akramian M, Khadivi A, Salehi-Arjmand H. Phenotypic and chemical variation of black mulberry (*Morus nigra*) genotypes. *Industrial Crops and Products*. 2018;117:260–271.
Hosseini SS, Khodaiyan F, Yarmand MS. Aqueous extraction of pectin from sour orange peel and its preliminary physicochemical properties. *International Journal of Biological Macromolecules*. 2016a;82:920–926.
Hosseini SS, Khodaiyan F, Yarmand MS. Optimization of microwave assisted extraction of pectin from sour orange peel and its physicochemical properties, *Carbohydrate Polymers*. 2016b;140:59–65.
Jafarzadeh-Moghaddam M, Shaddel R, Peighambardoust SH. Sugar beet pectin extracted by ultrasound or conventional heating: A comparison. *Journal of Food Science and Technology*. 2020;10.1007/s13197-020-04763-1.
Jamsazzadeh Kermani Z, Shpigelman A, Kyomugasho C, Van Buggenhout S, Ramezani M, Van Loey AM, et al. The impact of extraction with a chelating agent under acidic conditions on the cell wall polymers of mango peel. *Food Chemistry*. 2014;15:199–207.
Kazemi M, Khodaiyan F, Hosseini SS. Utilization of food processing wastes of eggplant as a high potential pectin source and characterization of extracted pectin. *Food Chemistry*. 2019a;294:339–346.
Kazemi M, Khodaiyan F, Hosseini SS, Najari Z. An integrated valorization of industrial waste of eggplant: simultaneous recovery of pectin, phenolics and sequential production of pullulan. *Waste Management*. 2019b;100:101–111.
Kazemi M, Khodaiyan F, Labbafi M, Hosseini SS, Hojjati M. Pistachio green hull pectin: Optimization of microwave-assisted extraction and evaluation of its physicochemical, structural and functional properties. *Food Chemistry*. 2019c;271:663–672.
Mesbahi GR, Jamalian J, Farahnaky A. A comparative study on functional properties of beet and citrus pectins in food systems. *Food Hydrocolloids*. 2005;19:731–738.
Mohammadzadeh J, Sadeghi-Mahoonak AR, Yoghbani M, Aalami M. Extraction of pectin from sunflower head residues of selected Iranian cultivars. *World Applied Sciences Journal*. 2010;8:21–24.
Pasandide B, Khodaiyan F, Mousavi Z, Hosseini SS. Pectin extraction from citron peel: optimization by Box–Behnken response surface design. *Food Science and Biotechnology*. 2018;27:997–1005.
Rahmani Z, Khodaiyan F, Kazemi M, Sharifan A. Optimization of microwave assisted extraction and structural characterization of pectin from sweet lemon peel. *International Journal of Biological Macromolecules*. 2020;147:1107–1115.
Raji Z, Khodaiyan F, Rezaei K, Kiani H, Hosseini SS. Extraction optimization and physicochemical properties of pectin from melon peel. *International Journal of Biological Macromolecules*. 2017;98:709–716.
Saberian H, Hamidi-Esfahani Z, Gavlighi HA, Barzegar M. Optimization of pectin extraction from orange juice waste assisted by ohmic heating. *Chemical Engineering and Processing: Process Intensification*. 2017;117:154–161.
Sahari MA, Akbarian M, Hamedi A. Effect of variety and acid washing method on extraction yield and quality of sunflower head pectin. *Food Chemistry*. 2003;83:43–47.
Samavati V. Polysaccharide extraction from *Abelmoschus esculentus*: Optimization by response surface methodology. *Carbohydrate Polymers*. 2013;95:588–597.
Samavati V, Yarmand MS. Statistical modeling of process parameters for the recovery of polysaccharide from *Morus alba* leaf. *Carbohydrate Polymers*. 2013;98:793–806.

Table 1.
A collection of underlined Iran-affiliated studies for the reported results about pectin extraction procedures from Agri-food wastes.

Pectin sources	Extraction conditions								Yield (%)	DM (%)	GalA (%)	MW (kg/mol)	Reference*
	Treatment	Solvent	Temperature (°C)	pH	S/L	Time	Power	Enzyme					
Beet pulp	CHE	HCl	80	1	1:50	3 h	—	—	20.0	58.92	66.18	116	Mesbahi et al. (2005)
Citron peels	CHE	citric acid	95	1.5	1:30	95 min	—	—	28.31	51.33	—	—	Pasandide et al. (2018)
Eggplant peel	CHE	citric acid	90	2.5	1:40	90 min	—	—	26.1	60.2	69.7	—	Kazemi et al. (2019b)
	UAE	citric acid	—	1.5	1:20	30 min	50 W		33.64	61.2	66.08		Kazemi et al. (2019d)
Grapefruit	UAE	HCl	70	1.5	1:50	25 min	—		17.92	75.1	68.21	68.3	Bagherian et al. (2011)
	MAE	HCl	—	—	1:50	6 min	900 W		27.81	80	75	50	Rahmani et al. (2020)
Pistachio green hull	MAE	—		1.5	1:15	165 s	700 W	—	18.13	12.1	66.0	1659	Kazemi et al. (2019c)
Sweet lemon peel	MAE	citric acid	—	1.5	—	3 min	700 W		25.31	5.80	87.2	615.8	Rahmani et al. (2020)
Sour orange peel	MAE	citric acid	—	1.50	1:15	3 min	700 W		28.8	1.5	71.0	—	Hosseini et al. (2016b)

*S/L: solid-liquid ratio, DM: degree of methyl esterification, GalA: galacturonic acid content, MW: molecular weight, UAE: U-assisted extraction, MAE: microwave assisted extraction, and CHE: conventional heating extraction. *see **Table 1**.*

Table 2.
The experimental profiles of pectin extraction for some Iran's agricultural wastes [9].

Author details

Roshanak Vakili
Department of Medical Library and Information Sciences, School of Health Management and Information Sciences, Iran University of Medical Sciences, Tehran, Iran

*Address all correspondence to: paper.publish@yahoo.com

References

[1] Bagherian H, Zokaee Ashtiani F, Fouladitajar A, Mohtashamy M. Comparisons between conventional, microwave-and ultrasound-assisted methods for extraction of pectin from grapefruit. Chemical Engineering and Processing Process Intensification. 2011;**50**:1237-1243

[2] Jafari F, Khodaiyan F, Kiani H, Hosseini SS. Pectin from carrot pomace: Optimization of extraction and physicochemical properties. Carbohydrate Polymers. 2016;**157**:1315-1322

[3] Hosseini SS, Parastouei K, Khodaiyan F. Simultaneous extraction optimization and characterization of pectin and phenolics from sour cherry pomace. International Journal of Biological Macromolecules. 2020;**158**:911-921

[4] Gharibzahedi SMT, Smith B, Guo Y. Pectin extraction from common fig skin by different methods: The physicochemical, rheological, functional, and structural evaluations. International Journal of Biological Macromolecules. 2019;**136**:275-283

[5] Khodaiyan F, Parastouei K. Co-optimization of pectin and polyphenols extraction from black mulberry pomace using an eco-friendly technique: Simultaneous recovery and characterization of products. International Journal of Biological Macromolecules. 2020;**164**:1025-1036

[6] Hosseini SS, Khodaiyan F, Kazemi M, Najari Z. Optimization and characterization of pectin extracted from sour orange peel by ultrasound assisted method. International Journal of Biological Macromolecules. 2019;**125**:621-629

[7] Nouri M, Mokhtarian M. Optimization of pectin extractions from walnut green husks and characterization of the extraction physicochemical and rheological properties. Nutrition and Food Sciences Research. 2020;7:47-58

[8] Vaez S, Karimi K, Mirmohamadsadeghi S, Jeihanipour A. An optimal biorefinery development for pectin and biofuels production from orange wastes without enzyme consumption. Process Safety and Environment Protection. 2020;**152**:513-526

[9] Belkheiri A, Forouhar A, Ursu AV, Dubessay P, Pierre G, Delattre C, et al. Extraction, characterization, and applications of pectins from plant by-products. Applied Sciences. 2021;**11**:6596

[10] Farsi M, Kalantar M, Zeinalabedini M, Vazifeshenas MR. First assessment of Iranian pomegranate germplasm using targeted metabolites and morphological traits to develop the core collection and modeling of the current and future spatial distribution under climate change conditions. bioRxiv. 2022; 10.1101/2022.03.14.484215

[11] Pant P, Pandey S, Dall'Acqua S. The influence of environmental conditions on secondary metabolites in medicinal plants: A literature review. Chemistry & Biodiversity. 2021. DOI: 10.1002/cbdv.202100345

[12] Li Y, Kong D, Fu Y, Sussman MR, Wu H. The effect of developmental and environmental factors on secondary metabolites in medicinal plants. Plant Physiology and Biochemistry. 2020;**148**:80-89

[13] Pang Z, Chen J, Wang T, Gao C, Li Z, Guo L, et al. Linking plant secondary

metabolites and plant microbiomes: A review. Frontiers in Plant Science. 2021;**12**:621276

[14] Hu L, Wu Z, Robert CAM, Ouyang X, Zust T, Mestrot A, et al. Soil chemistry determines whether defensive plant secondary metabolites promote or suppress herbivore growth. Proceedings of the National Academy of Sciences of the United States of America. 2021;**118**. DOI: 10.1073/pnas.2109602118

[15] Jan R, Asaf S, Numan M, Kim L, Kim KM. Plant secondary metabolite biosynthesis and transcriptional regulation in response to biotic and abiotic stress conditions. Agronomy. 2021;**11**:968

[16] Kashani A, Hasani M, Nateghi L, Asadollahzadeh MJ, Kashani P. Optimization of the conditions of process of production of pectin extracted from the waste of potato peel. Iranian Journal of Chemistry & Chemical Engineering. 2022;**41**:1288-1304

[17] Kazemi M, Amiri Samani S, Ezzati S, Khodaiyan F, Hosseini SS, Jafari M. High-quality pectin from cantaloupe waste: Eco-friendly extraction process, optimization, characterization and bioactivity measurements. Journal of the Science of Food and Agriculture. 2021;**101**:6552-6562

[18] Peighambardoust SH, Jafarzadeh-Moghaddam M, Pateiro M, Lorenzo JM, Domínguez R. Physicochemical, thermal and rheological properties of pectin extracted from sugar beet pulp using subcritical water extraction process. Molecules. 2021;**26**:1413

[19] Ezzati S, Ayaseh A, Ghanbarzadeh B, Khakbaz HM. Pectin from sunflower by-product: Optimization of ultrasound-assisted extraction, characterization, and functional analysis. International Journal of Biological Macromolecules. 2020;**165**:776-786

[20] Kazemi M, Khodaiyan F, Labbafi M, Hosseini SS. Ultrasonic and heating extraction of pistachio by-product pectin: Physicochemical, structural characterization and functional measurement. Journal of Food Measurement and Characterization. 2019;**14**:679-693

[21] Kazemi M, Khodaiyan F, Hosseini SS. Eggplant peel as a high potential source of high methylated pectin: Ultrasonic extraction optimization and characterization. LWT- Food Science and Technology. 2019;**105**:182-189

[22] Ramos-Aguilar OP, Ornelas-Paz JDJ, Ruiz-Cruz S, Zamudio-Flores PB, Cervantes-Paz B, Gardea-Béjar AA, et al. Effect of ripening and heat processing on the physicochemical and rheological properties of pepper pectins. Carbohydrate Polymers. 2015;**115**:112-121

[23] Santos JDG, Espeleta AF, Branco A, De Assis SA. Aqueous extraction of pectin from sisal waste. Carbohydrate Polymers. 2013;**92**:1997-2001

[24] Yapo BM, Wathelet B, Paquot M. Comparison of alcohol precipitation and membrane filtration effects on sugar beet pulp pectin chemical features and surface properties. Food Hydrocolloids. 2007;**21**:245-255

Chapter 4

Root and Tuber Crops: An Underexploited Source of Pectin and Future Prospects

Nneka R. Okereke and Chukwuemeka K. Nkere

Abstract

Starchy root and tuber crops are important global sources of carbohydrates with potentials to lift millions of people out of poverty across developing countries. The billion dollar pectin market which relies heavily on pectin isolated from fruits provides ample opportunities for non-conventional sources from the root and tuber food group. Pectins are abundant in higher plants and poses varying properties that can be used across industries including the food, the health and pharmaceutical sector, and in packaging regimes. We review current research into the isolation, modification, characterization and application of pectin sourced from root and tubers and explore the implications for an under-explored market in Africa. Despite the limited research conducted on root and tuber pectin, Citric acid used in the solvent method has shown to be a promising method of extraction, producing high pectin yields with industrial and pharmaceutical properties.

Keywords: root and tuber, food, pectin, isolation, Africa

1. Introduction

Before the introduction of cereals in many regions and especially in the tropics, root and tubers served as the only staple crop group that fed populations. Lebot [1] also described them as among the oldest crops on earth. Examples of these crops include, cassava, yam, potato, sweet potato and cocoyam. They share common characteristics including: vegetative propagation, breeding approaches, post-harvest issues (bulky and perishable), requiring low inputs, adaptability to mixed farming and heavily involves women throughout their production value chains [2].

These food security crops serve as major sources of calories, nutrition, income and employment for millions of people, especially across developing countries in Sub-Saharan Africa, Asia and Latin America [3]. According to current FAO estimates, roots and tubers are the fourth most produced group of crops after cereals, sugar crops and fruits (**Figure 1**) and are the second most important food group, serving around 300 million people across developing countries [2, 4]. Africa produces the highest root and tuber crops at an estimated value of over 300 million tonnes across approximately 39 million hectares of land (**Table 1**). In this region, the food group

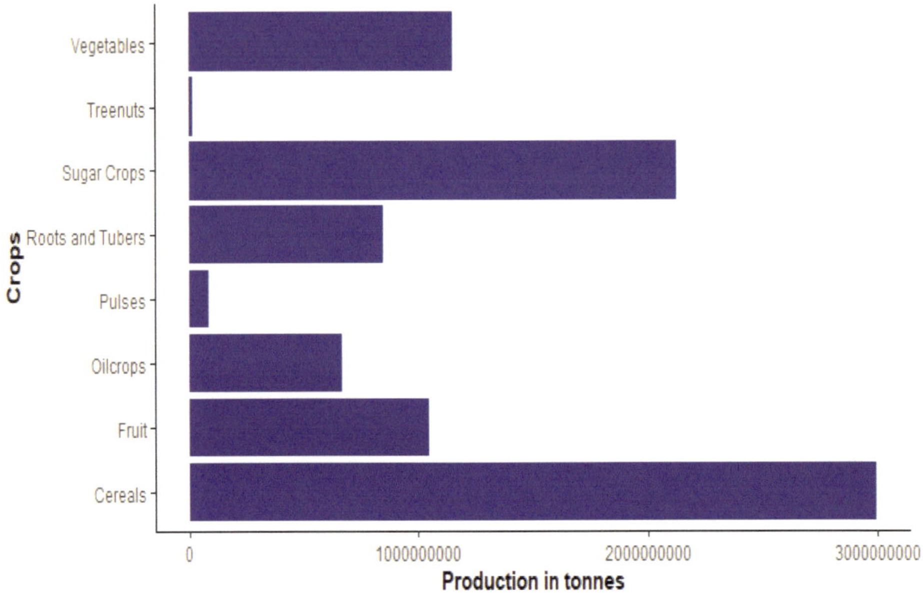

Figure 1.
2020 FAO production estimates of crop groups produced around the world.

	Africa	America	Asia	Europe	Oceania
Production (t)	333,594,726	78,182,331	324,104,149	107,693,897	4,047,282
Area harvested (ha)	38,953,062	4,104,037	15,590,781	4,571,456	318,331

Table 1.
Estimated production and area harvested values of root and tuber crops distributed around the world.

plays an integral part of social cultural activities [5] and has shown great potentials in alleviating poverty and improving the resilience of its mostly limited resourced stakeholders in tackling the growing threats from climate change [6].

For these reasons, root and tubers have received, in the past 20 years, invaluable attention and research funding from national and international agencies like the Consultative Group on International Agricultural Research (CGIAR), the Bill and Melinda Gates Foundations (BMGF), financial institutions (e.g. World Bank) and National Agricultural Centres [7]. These concerted efforts have resulted in the increase in production (development of improved varieties and formal seed quality systems) and innovative research into the complex roles this particular food group plays in nutrition, as animal feed, as raw materials for starch and alcohol production, and fermented foods and beverages [8–10].

2. Pectin research in root and tubers

Root and tubers store edible starchy materials in their stems, roots, rhizomes, corms, and tubers and serve as sources of important bioactive compounds and polymers with nutritional and pharmaceutical potentials [4]. One of the most important but least explored polymers in root and tuber crops are pectins. In higher plants, pectins are an integral part of the primary cell wall and middle lamella and are also most abundant in non-woody parts of the plant [11]. They are non-starch heteropolysaccharides composed mainly of covalently α-1,4-linked D-galacturonic acid (GalA) units (peculiar to the homogalacturonan pectin chain) with the urinate residues in their natural state partially esterified. But they may vary in chemical composition, sugar content and molecular weight based on the development stage of the plant, isolation condition, storage and process of production [12]. Pectins can also be classified based on their degree of esterification and chemical structure. These characteristics contribute to the multi-functionality of pectin that include: textural and rheological properties of plant-based foods, movement of water and nutrients, maintaining the turgidity and concentration of the cell wall and their applications in food and industry [13, 14]. In food and pharmaceutical industries these polymers have been used, for years, as gelling agents, emulsifiers, stabilizers and thickeners [15–17]. Therapeutically, pectins are soluble dietary fibers, can reduce blood cholesterol, used in drug delivery, replace fats in confectionery and improve pre-biotic activities in the gut [11].

Although abundant in most higher plants, commercially produced pectins are primarily sourced from fruit residue (citrus peels and apple pomace especially) generated in large quantities from fruit processing companies. In this chapter, we will be reviewing the structure, extraction, modification and application of pectins from root and tubers and exploring the impact of the food group as a non-conventional source of commercial pectin in Africa.

2.1 Pectin extraction

Extraction techniques are a major factor in determining the quality and yield extracted from pectin sources. Some of which include the most common - microwave heating and direct boiling using chemicals [18]; and the more novel methods that include the use of ultrasonic sound, autoclave, enzymes and electromagnetic induction [19]. In some cases methods have been combined to extract pectin [20]. Pectin from root and tuber crops as well as their food products and by-products have been extracted using different extraction methods (**Table 2**). Infante et al. [21] reported the extraction of pectin from cassava bread using a boiling solution of oxalic acid and ammonium oxalate, while Coelho et al. [22] extracted pectin from cassava residue using ammonium oxalate and further precipitated with ethanol. Menoli and Belia [23] also reported the precipitation of pectin from cassava galacturonic acid using a modified method [30] that involved saturated potassium acetate solution and 95% ethanol in a 4:1 ratio. Ogutu and Mu [24] before evaluating the effect of ultrasonic factors on sweet potato pectin extracted the polymer by using sodium hexametaphosphate. Ultrasound and microwave assisted acid extraction methods have also been used to extract pectin from sweet potato residues [25]. Citric acid was reported as an ideal solvent for the extraction of pectin from sweet potato peels [26]. Yang et al. [27] extracted higher yields of pectin than sourced from citrus and apple, using different acidic solvents from potato pulp. For yam pectin, Tang et al. [28] recently isolated the polymer from *Dioscorea opposita* tubers using the enzyme extraction method while Effah-Manu et al. [29] successfully extracted pectin from *D. rotundata* and *D. alata* tubers using citrus acid as a solvent.

Pectin source	Method of extraction	Characterization	Application	References
Cassava bread	**Solvent extraction**: Extracted with a boiling solution of 0.25% oxalic acid and 0.25% ammonium oxalate **Yield**: 0.34–0.61%	No	No	[21]
Cassava residue	**Solvent extraction**: Three volumes 0.5% (w/v) ammonium oxalate solution and ethanol (92.8° GL) **Yield**: 0.1–0.5%	No	Used to coat nanoparticles containing β-carotene aiming at the gastrointestinal administration of this lipophilic nutraceutical.	[22]
Cassava galactoronic acid	**Solvent extraction**: 5 ml cooking liquor, three volumes of saturated potassium acetates olution and 95% ethanol (4:1)	No	No	[23]
Sweet potato galacturonic acid	**Solvent extraction**: Using sodium hexametaphosphate, purified using an ultrafiltration column then precipitated and washed successively with 60%, 75% and 90% ethanol	12% degree of methoxylation (DE) I.e. low methoxy pectin	No	[24]
Sweet potato residue	Ultrasound/microwave assisted acid extraction method: extraction by HCl (pH 2), centrifuged and collected supernatant was purified through ultrafiltration. Pectin was precipitated using three volumes (v/v) of absolute ethanol and further washed three times with ethanol at different concentrations	Molecular weight: 7.53 × 105 g mol-1 DE: 33.17%; low methoxy pectin	Modified pectin with improved emulsifying properties	[25]
Sweet potato residue	Solvent extraction: using citric acid at different temperatures (60,80,100°C), for different times (40,70 and 100mins) under different pH (1.0,1.5 and 2.0) Yield: 1.9–64%	DE: 58.5%; high methylation pectin	High emulsifying properties	[26]

Pectin source	Method of extraction	Characterization	Application	References
Potato pulp	Solvent extraction: Using HCl, H_2SO_4, HNO_3, citric acid, and acetic acid. Yield: Citric acid produced the highest yield at 14.34%	highly branched rhamnogalacturonan I domain DE: 37.45%; low methoxy pectin	High emulsifying activity and emulsion stability	[27]
Yam residue (Chinease Yam Polysaccharides)	Extracted with pure water and treated with α-amylase (95°C, 90 min), glucoamylase (60°C, 30 min), papain (60°C, 40 min) and enzyme deactivation (100°C, > 10 min) in sequence. The compex was further precipitated with ethanol concentration of 80% (v/v).	33.2%; low methyl-esterified pectin ~38.1% highly branched rhamnogalacturonan I	No	[28]
Yam residue	Solvent extraction: using citric acid, washed and precipitated with ethanol (70, 80 and 90%). Yield: ranged from 4.32–15.88%	DE: 30.52–51.37%	Significantly contributes to the rhelogical and textural properties of prepared Dioscorea species	[29]

Table 2.
Extraction, characterization and application of non-conventional root and tuber pectin sources.

2.2 Characterization of root and tuber pectins

In addition to its characteristic α-1,4-linked D-galacturonic acid units, pectins are composed of chains which classify them into the most abundant classes, namely homogalacturonan and rhamnogalacturonan I [31, 32]. Minor components also include substituted galacturonans: rhamnogalacturonan II (most conserved structure), xylogalacturonan, and apiogalacturonan - mostly found in aquatic plankton. Pectins can also be characterized based on their degree of methylation [33]; where a value higher that 50% is described as a high methyl pectin and as a low methyl pectin when the value is lower than 50%. Although characterization of root and tuber pectins are limited, previous studies have reported that potato residue and sweet potato peels contain large quantities of rhamnogalacturonan I, which is the hairy region of pectin [34]. They also produce high methyl pectin indicating their emulsifying potentials (**Table 2**).

With a series of antibodies and enzymes, Staack et al. [35] identified a diversity of pectin structures in cassava extracts—including methyl-esterifed homogalacturonan and rhamnogalacturonan-II, implying its pre-biotic potentials. Yams are reported to have low methyl pectin [28, 30], although this varies with the spices.

2.3 Application of root and tuber pectin

One of the most important attributes of food products from root and tubers, which determines its acceptability, is its textural and rheological properties [36]. Although these properties are mostly determined by their starchy nature, high pectin yields with high methyl values have been especially reported in potato and sweet potatoes with demonstrable links between pectin structure and the textural properties of their food products [3]. Studies showing this link include the evaluation of cassava and cocoyam root softening and implicates the role of endogenous (pectin methylestherase) and microbial enzymes which target cell wall materials [37, 38]. To improve the bioavailability of beta carotene in animal feed, nano-particles containing the molecule was coated with pectin sourced from cassava residue [22]. The research showed promising strategy for neutraceutical administration of important bioactive compounds through mucosal surfaces.

3. Conclusion and future prospects

From current research, root and tubers have shown potentials as unconventional sources of commercial pectin. They have shown promising emulsifying, textural, pharmaceuticals and neutraceutical properties that can be applied to a variety of industries.

These starchy crops already serve as raw materials for several markets and industries but can compete favorably in the current commercial pectin market which has citrus fruits as its highest contributor. The current pectin market is valued at US$994.45 million and it is expected to exceed US $1 billion by 2030. Africa being the highest producer of root and tubers is poised to be a benefactor from this billion dollar market. It is imperative that national and international agencies fund research into improving root and tuber pectin yields, and structures by optimizing extraction and modification techniques.

Author details

Nneka R. Okereke and Chukwuemeka K. Nkere*
National Root Crops Research Institute Umudike, Umuahia, Abia State, Nigeria

*Address all correspondence to: nkerechukwuemeka@yahoo.com

References

[1] Lebot V. Tropical root and tuber crops cassava, sweet potato, yams and aroids. UK.CABI.279-434. 2009

[2] Thiele G, Khan A, Heider B, Kroschel J, Harahagazwe D, Andrade M, et al. Roots, tubers and bananas: Planning and research for climate resilience. Open Agriculture. 2017;**2**(1):350-361

[3] Harris KP, Goddard JR, Kelly A, Cullen A, Reynolds T, Anderson CL. Root, tuber, and Banana textural traits. EPAR. 2015;**2015**:1-52

[4] Chandrasekara A, Josheph Kumar T. Roots and tuber crops as functional foods: A review on phytochemical constituents and their potential health benefits. International Journal of Food Science. 2016;**2016**:3631647. DOI: 10.1155/2016/3631647

[5] Obidiegwu JE, Lyons JB, Chilaka CA. The Dioscorea genus (yam)—An appraisal of nutritional and therapeutic potentials. Food. 2020;**9**:1-45

[6] Thiele G, Friedmann M, Polar V, Campos H. Overview. In: Thiele G, Friedmann M, Campos H, Polar V, Bentley JW, editors. Root, Tuber and Banana Food System Innovations. Cham: Springer; 2022. DOI: 10.1007/978-3-030-92022-7_1

[7] Nteranya S, Adiel M. Root and Tuber Crops (Cassava, Yam, Potato and Sweet potato). In: Paper Presented at an International Conference in Senegal. 2015. pp. 1-26

[8] Moyo M et al. Orange-fleshed Sweetpotato puree: A breakthrough product for the bakery sector in Africa. In: Thiele G, Friedmann M, Campos H, Polar V, Bentley JW, editors. Root, Tuber and Banana Food System Innovations. Cham: Springer; 2022. DOI: 10.1007/978-3-030-92022-7_5

[9] Scott GJ, Best R, Rosegrant M, Bokanga M. Roots and tubers in the global food system: A vision statement to the year 2020 (including annex). A co-publication of the international potato Centre (CIP). In: Centro Internacional de Agricultural Tropical (CIAT), International Food Policy Research Institute (IFPRI), International Institute of Tropical Agriculture (IITA), and International Plant Genetic Resources Institute (IPGRI). Printed in Lima. Peru: International Potato Centre; 2000

[10] Ugwu FW. The potentials of root and tubers as weaning foods. Pakistan Journal of Nutrition. 2009;**8**(10):1701-1705

[11] Srivastava P, Malviya R. Sources of pectin, extraction and its applications in pharmaceutical industry – an overview. Indian Journal of Natural Products and Resources. 2011;**2**:10-18

[12] Novoselskaya IL, Voropaeva NL, Semenova LN, Rashidova SS. Trends in the science and applications of pectins. Chemistry of Natural Compounds. 2000;**36**:1-10

[13] Freitas CMP, Coimbra JSR, Souza VGL, Sousa RCS. Structure and applications of pectin in food, biomedical, and pharmaceutical industry: A review. Coatings. 2021;**11**:922

[14] Wang W, Chen W, Zou M, Lv R, Wang D, Hou F, et al. Applications of power ultrasound in oriented modification and degradation of pectin: A review. Journal of Food Engineering. 2018;**234**:98-107

[15] Min B, Bae IY, Lee HG, Yoo SH, Lee S. Utilization of pectin-enriched materials from apple pomace as a fat replacer in a model food system. Bioresource Technology. 2010;**101**(14):5414-5418

[16] Nakamura A, Furuta H, Maeda H, Takao T, Nagamatsu Y. Structural studies by stepwise enzymatic degradation of the main backbone of soybean soluble polysaccharides consisting of galacturonan and rhamnogalacturonan. Bioscience, Biotechnology, and Biochemistry. 2002;**66**(6):1301-1313

[17] Saha D, Bhattacharya S. Hydrocolloids as thickening and gelling agents in food: A critical review. Journal of Food Science and Technology. 2010;**47**(6):587-597

[18] Maxwell EG, Belshaw NJ, Waldron KW, Morris VJ. Pectin – An emerging new bioactive food polysaccharide. Trends in Food Science and Technology. 2012;**24**:64-73

[19] Chandel V, Biswas D, Roy S, Vaidya D, Verma A, Gupta A. Current advancements in pectin: Extraction. Properties and Multifunctional Applications. Foods. 2022;**11**:2683. DOI: 10.3390/foods11172683

[20] Yang Y, Wang Z, Hu D, Xiao K, Wu JY. Efficient extraction of pectin from sisal waste by combined enzymatic and ultrasonic process. Food Hydrocolloids. 2018a;**79**:189-196

[21] Infante RB, Garcia OO, Rivera C. Characterization of dietary fiber and pectin of cassava bread obtained from different regions of Venezuela. Revista Chilena de Nutricion. 2013;**40**(2):169-173. DOI: 10.4067/S0717-75182013000200012

[22] Coelho B, Mazzarino L, Pitz HS, Feltrin C, Voytena APL, Coelho DS, et al. Development of nanoparticles coated with cassava bagasse pectin (Manihot esculenta Crantz) containing β-carotene for mucoadhesive applications. Anais da Academia Brasileira de Ciencias. 2020;**92**(4):e20200134. DOI: 10.1590/0001-3765202020190134

[23] Menoli AV, Beleia A. Starch and pectin solubilization and texture modification during pre-cooking and cooking of cassava root (Manihot esculenta Crantz). Food Science and Technology. 2005;**40**(4):744-747

[24] Ogutu FO, Mu TH. Ultrasonic degradation of sweet potato pectin and its antioxidant activity. Ultrasonics Sonochemistry. 2017;**38**:726-734

[25] Arachchige MPM, Mu T, Ma M. Structural, physicochemical and emulsifying properties of sweet potato pectin treated by high hydrostatic pressure and/or pectinase: A comparative study. Journal of the Science of Food and Agriculture. 2020;**100**(13):4911-4920. DOI: 10.1002/jsfa.10552

[26] Hamidon NH, Abang Zaidel DN, Mohd Jusoh YM. Optimization of pectin extraction from sweet potato peels using citric acid and its emulsifying properties. Recent Patents on Food, Nutrition and Agriculture. 2020;**11**(3):202-210. DOI: 10.2174/2212798411666200207102051

[27] Yang JS, Mu TH, Ma MM. Extraction, structure, and emulsifying properties of pectin from potato pulp. Food Chemistry. 2018b;**244**:197-205. DOI: 10.1016/j.foodchem.2017.10.059

[28] Tang W, Liu D, Wang J-Q, Huang X-J, Yin J-Y, Geng F, et al. Isolation and structure characterization of a

low methyl-esterified pectin from the tuber of *Dioscorea opposita* Thunb. Food Chemistry. 2021;**2021**:35

[29] Effah-Manu L, Maziya-Dixon B, Wireko-Manu FD, Agbenorhevi JK, Oduro I. Yam pectin and textural characteristics: A preliminary study, International Journal of Food Properties. 2002;**25**(1):1591-1603. DOI: 10.1080/10942912.2022.2096065

[30] Selvendran RR, O'Neill MA. Isolation and analysis of cell walls from plant material. Methods of Biochemical Analysis. 1987;**32**:25-153

[31] Belkheiri A, Forouhar A, Ursu AV, Dubessay P, Pierre G, Delattre C, et al. Extraction, characterization, and applications of pectins from plant by-products. Applied Sciences. 2021;**11**:6596. DOI: 10.3390/app11146596

[32] Ishii T, Matsunaga T, Pellerin P, O'Neill MA, Darvill A, Albersheim P. The plant cell wall polysaccharide rhamnogalacturonan II self-assembles into a covalently cross-linked dimer. The Journal of Biological Chemistry. 1999;**274**(19):13098-13104

[33] Maric M, Grassino AN, Zhu Z, Barba FJ, Brnčic M, Brnčic SR. An overview of the traditional and innovative approaches for pectin extraction from plant food wastes and by-products: Ultrasound-, microwaves-, and enzyme-assisted extraction. Trends in Food Science and Technology. 2018;**76**:28-37

[34] Byg I, Diaz J, Øgendal LH, et al. Large-scale extraction of rhamnogalacturonan I from industrial potato waste. Food Chemistry. 2012;**131**:1207-1216

[35] Staack L, Della Pia EA, Jørgensen B, Pettersson D, Rangel Pedersen N. Cassava cell wall characterization and degradation by a multicomponent NSP-targeting enzyme (NSPase). Scientific Reports. 2019;**2019**:9

[36] Yuliarti O, Gusti E, Chiang JH, Teo PX, Ng JY. Rheological and microstructural properties of native cassava starch-low methoxyl pectin in a fruit filling gel system. LWT. 2021;**146**:111568. DOI: 10.1016/j.lwt.2021.111568

[37] Ampe F, Keleke S, Robert W, Brau A. The role and origin of pectin degrading enzymes during cassava retting. Transformation Alimentaire du Manioc. 1995;**1995**:333-344

[38] Mbouobda HD, Fotso TMA, Muyang RF, Omokolo ND. Improvement of cocoyam productivity with effective microorganisms (EM) and indigenous microorganisms (IMO) manures. American Journal of Biochemistry. 2015;**5**(2):42-47

Chapter 5

Applications of Pectinolytic Enzymes in Process Industries

Haneef Ur Rehman

Abstract

The manufacturing of industrial important products by using enzyme technology is sustainable method and much to offer as compared to using chemical catalyst. The enzyme can manage the industrial preparation under mild reaction conditions using specific substrate. Enzyme uses shorter time, produces limited or no wastes and eco-friendly in nature. Pectinase complex enzyme system catalyzes the breakdown of pectin polymers. Pectinase is a generic term used for a group of enzymes that catalyzes the degradation of pectin substances by hydrolysis, trans-elimination, as well as, de-esterification reactions. Pectinase is produced from various organisms including bacteria, fungi, yeast, insects, protozoa and nematodes. Microorganism is usually used for industrial production of pectinase because of its easily growth and cost-effective downstream process. Pectinase has various applications in different industrial process such as fruit juice extraction, treatment of wastewater, papermaking, degumming of plant-based fibers, coffee and tea fermentation. This chapter describes the importance of pectinase and its application in different industrial process. Furthermore, it gives detail review about the pectinase and opportunities for the future research.

Keywords: pectinase, pectin, industrial applications, pectinolytic enzymes, industrial processes

1. Introduction

1.1 Pectin

The cell wall surrounds the plant cell and protects the cell and its distinctive components that are essential for plant survival from different environmental pressure. The plants cell wall supports the plant to survive under various climate conditions and the polysaccharides in most abundant material composition of cell as compared to protein, aromatic and aliphatic compounds. Pectin polymer plays significant role in the composition of cells and acts as cementing agent to structural organization and functionality of cell. Pectin is mostly present in the cell walls of higher plants. It is one of the major components of plants and responsible for the

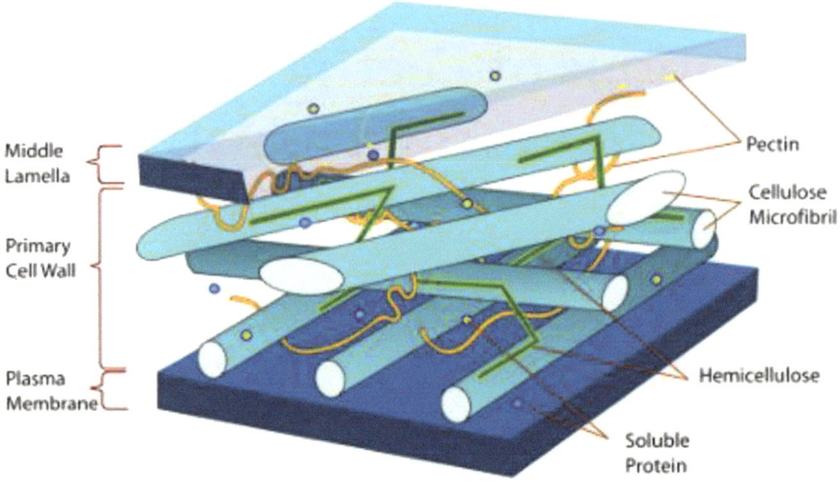

Figure 1.
Cell wall structure of higher plants [1].

structure integrity and cohesion of plant tissues (**Figure 1**) [2, 3]. The term pectin was initially quoted in eighteenth century from tamarind fruit as a peculiar substance [4]. The basic characterization of pectin was done on nineteenth century and categorized it as active component of fruit and fruit-based products [5]. As an outcome, Nussinovitch suggested the word 'pectin' in Ref. to Greek work 'pektikos' which mean 'congeal, solidify or curdle' [6]. The actual chemistry of pectin began when Ehrlich [7] discovered that the D- galacturonic, an isomer of D-gluronic acid, is the central constituent of pectin, and the some of these D-galacturonic acid are partially esterified with methyl alcohol. Now, pectin is thought to be composed of at least seventeen kinds of different monosaccharides, in which D-galacturonic acid is usually the most abundant, followed by D-galactose or L-arbinose [8, 9].

1.1.1 Structure of pectin

The pectin is a complex carbohydrate polymer of galacturonic acid residues having esterified carboxylic groups with methanol. The degree of esterification of carboxylic groups of the pectin has been changed from different sources. Pectin consisted of four polysaccharides including homogalacturonan, RGI (xylogalacturonan rhamnogalacturonan I) and RGII (rhamnogalacturonan II) (**Figure 2**) [11, 12]. The homogalacturonan is the main linear chain of galacturonic acids linked through α-1, 4 linked and makes the main chain of pectin polymer. RGI formed through the linkage of ragmnopyranose residues by α-1, 2 bond and makes the branches with the polymer. D-apiose, 2-O-methyl-D-xylose-and 2-O-methyl-L- fructose side chain generates the RGII region of pectin polymer [13]. The C-2 or C-3 position of galacturonic acid chain is acetylated in rhamnogalacturonan 1 region of pectin polymer, and most of the chains are consisted of D-galacturonic acids. In aqueous conditions the pectin polymer stand in pure form conformation with great flexibility and do not form straight conformation. This composition of pectin was varied among different sources [14]. The properties of pectin are strongly depending on methylation of galacturonic acids residues, which is usually 70%.

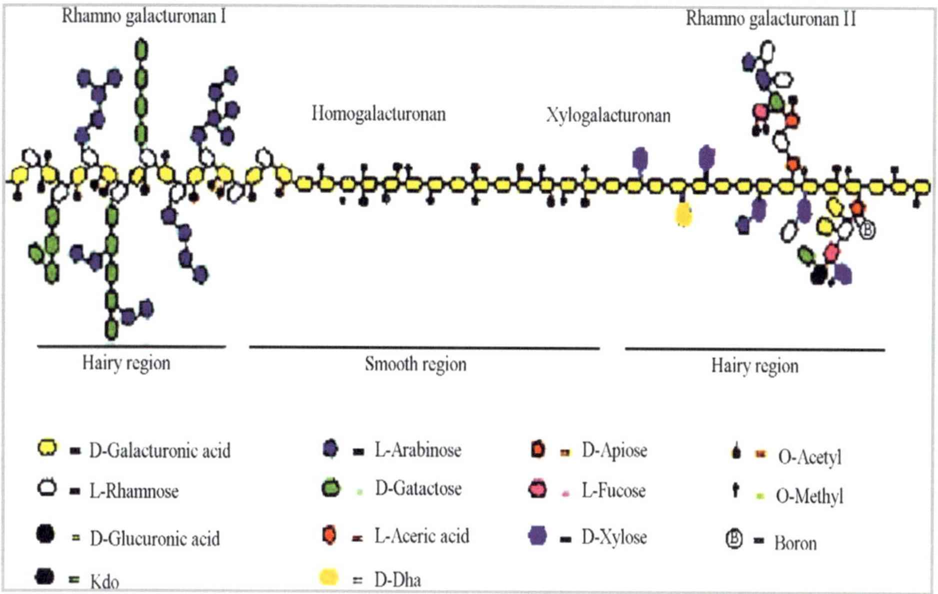

Figure 2.
Basic structure of pectin [10].

1.1.2 Classification of pectin

The pectin polymer is divided into protopectin, pectic acid, pectinic acid and pectin because of variation in backbone chain of galacturonic acid residues [15].

1.1.2.1 Protopectin

Protopectin term is used to illustrate the insoluble pectin. It is parent pectin and yield pectin or pectinic acid on restricted hydrolysis.

1.1.2.2 Pectic acid

Pectic acid contains insignificant amounts of methoxy groups. It is soluble form of pectin.

1.1.2.3 Pectinic acid

Pectinic acid is the polygalacturonan that contains various amounts of methoxy groups having the property to form gel with sugar and acid.

1.1.2.4 Pectin (Polymethylgalacturonate)

Pectin is the polygalacturonic acid in which the galacturonic acid residues are 75% methylated.

1.1.3 Distribution of pectin in nature

Pectin is widely disturbed in nature and mostly found in angiosperms and gymnosperms plants along with pteridophytes, bryophytes, lycophytes and carophytes [16]. The pectin represents 35% of total plant biomass and among the high percentage containing biomolecules in plants [17]. The concentration of pectin in plants varied according to the plant types, maturation time and its environmental conditions (**Table 1**). The 100 grams of citrus peels and apple pomaces contained 20 grams of pectin and have been used for commercial extraction of pectin [21].

1.1.4 Commercial utilization of pectin

Pectin has been used in various industrial processes and applications due to its biocompatibility, cost-effective, easily available and among the most abundant natural compounds on earth. The food additive, thickening agent, gelling agent, cosmetic texturizing agent, a good source of dietary fiber and also components of biodegradable films, adhesives, paper and antimicrobial food packaging are the major applications of

Fruits	Botanical name	Pectin content (%)	References
Apple	*Malus* spp	0.5–1.6	[18]
Apple pomace		1.5–2.5	[19]
Banana	*Musa acuminate* L.	0.7–1.2	[18]
Beet pulp	*Beta vulgaris*	1.0	[19]
Carrot	*Daucus carota*	0.2–0.5	[19]
Giant granadilla	*Passiflora quandrangularis* L.	0.4	[20]
Carambola	*Averrhoa carambola*	0.66	[20]
Guava	*Psidium guajava* L.	0.77–0.99	[20]
Lemon pulp	*Citrus limon*	2.5–4.0	[19]
Lychee	*Litchi chinesis* S.	0.42	[18]
Mango	*Mangifera indica* L.	0.26–0.42	[20]
Orange peel	*Citrus sinesis*	3.5–5.5	[19]
Papaya	*Caricia papaya*	0.66–1.0	[20]
Passion fruit	*Passiflora edulis* L.		[20]
Passion fruit rind		2.1–3.0	[20]
Peaches	*Prunus persica*	0.1–0.9	[18]
Pineapple	*Ananas comosus* L.	0.04–0.13	[20]
Strawberries	*Fragaria ananassa*	0.6–0.7	[20]
Tamarind	*Tamarindus indica* L.	1.71	[20]
Thimbleberry	*Rubus rosalfolius*	0.72	[20]
Tomato fruit	*Lycopersicon esculentum*	0.2–0.6	[18]

Table 1.
Pectin content in various fruits.

pectin in industries [22]. The mechanism of enzymatic modification of pectin polymer needs to be understood to further increase the industrial scope of pectin polymer.

1.2 Pectinase

Pectinase is a complex group of enzyme that catalyzes the degradation of pectin polymer [23]. Pectinase has wide range of applications in fruit juices preparation, textile processing, papermaking, pectin containing wastewater treatment, degumming of plant's bast fibers, wine clarification, oil processing and coffee and tea industry.

1.2.1 Classification of pectinase

On the basis of pectin degrading mechanism, the pectinase enzymes are classified into polygalacturonase, pectin lyase and pectin esterase (**Figure 3**).

1.2.1.1 Polygalacturonase

Polygalacturonase catalyzes the hydrolysis of pectin polymer into D- galacturonic acid monomer by addition of water molecules in α-1, 4 glycosidic linkages. Polygalacturonase is the most extensively studied enzyme among different pectinolytic enzymes and further classified into exo-polygalacturonase and endo-polygalacturonase by breaking of external and internal glycosidic bond of polymer chain, respectively.

1.2.1.2 Pectin lyase

Pectin lyase catalyzes the breaking of α- 1, 4 glycosidic bonds and generates galacturonide with unsaturated galacturonic acid. Pectin lyase is classified into

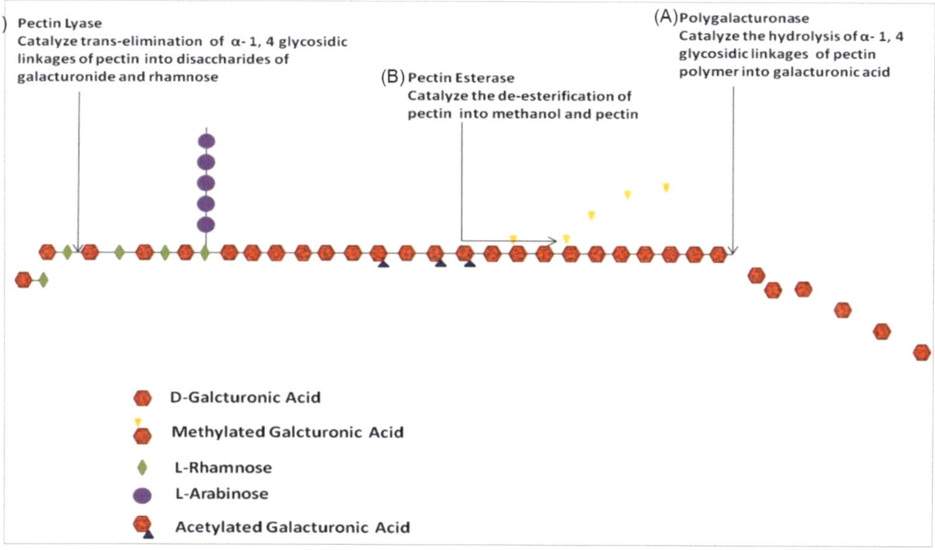

Figure 3.
Mode of degradation of pectin by different pectinases. (A) Polygalacturonase (B) pectin esterase (C) pectin lyase [24].

exo-pectin lyase and endo-pectin lyase on the basis of catalyzing the breaking of α- 1, 4 glycosidic bonds sequentially and randomly, respectively.

1.2.1.3 Pectin esterase

Pectin esterase is also known as pectin methyl hydrolase that catalyzes the de-esterification of pectin molecule into pectic acid and methanol.

1.2.2 Structure of pectinase

The structure of pectinase (**Figure 4**) enables to understand the enzyme reaction mechanism on molecular basis and also provide information about the structural differences between enzymes that directly affect the catalytic properties of enzymes. Yoder *et al.* [26] characterized the three-dimensional structure of pectinase from *E. chrysanthemis*. The structure of pectinases composed of a domain of parallel β strands folded into a large right-handed cylinder. The inner cylinder contains seven to nine helical turns and due to distinctive arrangement of three β- strands in each turn of helix look like a prism shaped. The consecutive stands assemble to form three parallel β-sheets named PB1, PB2 and PB3. Even the mechanism of pectin degradation by three pectinases is different, the substrate binding figure out from structure, sequence similarity and site directed mutagenesis studies [27], are all observed in a similar location within cleft formed on the outer of the parallel β- helix among one side of PB1 and the overhanging loops [28].

1.2.3 Sources of pectinase

Pectinase plays a very significant role in various biological processes across the entire field of living organism. The pectinase is widely distributed in nature and produced by different living organisms such as plants, microorganisms, insects, nematodes and protozoa. Microorganisms are considered as primary source for the production of industrial important enzymes.

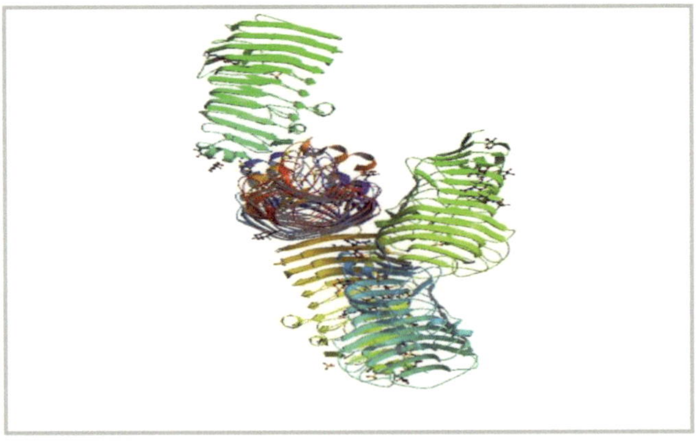

Figure 4.
Three-dimensional structure of pectinase [25].

1.2.3.1 Microbial pectinase

Pectinase is one of the important factors in the plant pathological process, plant-microbe symbiosis and in the decomposition of plant decay matters. The microbial pectinase is playing important role in nature by contributing natural recycling of carbon in the environment. Different microorganisms are known to produced pectinase with different molecular mass and catalytic properties. *Aspergillus niger* is mostly used for the industrial production of pectinase [29]. The pectinases from fungus sources are usually acidic in nature and only can work in acid conditions. Production of alkaline pectinase remains under developed, as only few reports are available on the production alkaline pectinase by bacterial strains [30–32]. Alkaline pectinase can be used for the treatment of pectineus substances containing wastewater from vegetable and food processing industries [33].

1.2.3.2 Bacillus

Bacillus is one of the large genera of bacterial strains. It is rod-shaped, endospore bearing bacteria and belongs to the family *Firmicutes.* The endospores of *Bacillus* are more resistant to heat, drying, disinfectants and other destructive agents. The genus *Bacillus* covered a great diversity of strains and some of them strictly aerobic, while the other are facultative anaerobic. The *Bacillus* especially *Bacillus subtilis* and *Bacillus licheniformis* are excellent candidates for large-scale production of commercially important enzymes.

1.2.3.3 B. licheniformis

B. licheniformis is a gram positive, rod-shaped and endospore forming bacterial strain, used for the production of different industrial important products. *B. licheniformis* is a saprophytic bacterium in nature and commonly found in soil and other natural environment. The *B. licheniformis* is capable of growing on a large diversity of nutrient sources because of synthesizing and secreting different hydrolytic enzymes, and this quality makes the *B. licheniformis* an industrial important microorganism.

Following are some of very important commercial applications of *B. licheniformis.*

1. *B. licheniformis* is widely used for production of commercially important thermostable enzymes like protease and α-amylase which are stable at 105–110°C for short period of times [34].

2. *B. licheniformis* is also used to produce commercially important antibiotics such as bacitracin and surfactin, as well as poly-gamma-glutamic acid in great numbers [34].

3. Thermostable α–amylase from *B. licheniformis* has been used for the liquefaction of wheat flour [35] and corn meal [36], and the hydrolysate was then saccharified to produce ethanol using *Saccharomyces cerevisiae*.

1.2.4 Production of pectinase

Fermentation technology has been effectively used in pectinase production by both fungus as well as bacterial strains. On the basis of production of pectinase, SSF fermentation provides higher productivity as compared to SmF fermentation [37]. Even with the advantage of high productivity, the industrial application of SSF is hard to visualize due to difficultly in product recovery. SmF is well developed fermentation technique used for large-scale production of metabolites and technically easier to perform as compared to SSF [38].

1.2.5 Biochemical properties of pectinase

Biochemical properties of pectinase are very important for their commercialization on industrial scale. The characteristics of pectinase produced from different microorganisms have been reported, and the biochemical properties of enzyme were varied from source to source [39–41]. The pectinase from various microorganism has different range of optimum temperature (30–60°C) and pH (3.0–9.0) for maximum enzymatic activities [40]. They also have different molecular weight, thermal stability and kinetic parameters [11]. Most of the pectinase have been reported to perform maximum activity in range of 40–50°C [42–44]. Mohamed *et al.* [45] reported that pectinase from *Trichoderma harzianum* showed 100% stability at 30°C after 60 minutes of incubation. The stability of enzyme provides valuable information about its structure and function. *Bacillus* sp. MG-cp-2 produced alkaline pectinase having stability at broad pH range and retained 80% of its initial activity at room temperature after 24 hours [46]. The activity of pectinase was increased in the presence of Fe^{2+}, Mn^{2+}, Ca^{2+} and Cu^{2+}, but strongly inhibited by Mg^{2+} [41]. Hg^{2+}, Zn^{2+} and Cu^{2+} were found to inhibit the pectinase activity, while Mn^{2+} and Co^{2+} enhanced the pectinase activity from *S. sclerotinium* [47].

1.2.6 Applications of pectinase

Pectinase is one of the very important enzymes in fruits and textiles industries as well as having different biotechnological application in various industries. Following are some important applications of pectinase in different aspects.

1.2.6.1 Fruit and vegetable juices industries

Pectinase has been widely used in fruit and vegetable juices industries. These industries commercially produced different juices including sparkling clear juices, cloudy juices and unicellular product. In sparkling clear juices, pectinase is added to enhance the production juices by hydrolyzing pectin particles [48].

1.2.6.2 Textile industry

Pectinase with the combination of other enzymes such as amylase, lipase, cellulase and hemicellulose has been used to degrade sizing agents from cotton in textile industry. Uses of enzymes ensured the low discharge of waste chemical in environment and improved both the environmental safety and the value of product. Traditionally the scouring of cotton was done by using 3–6% aqueous sodium hydroxide and high energy. Bioscouring is a novel process of using enzymes to specifically remove the non-cellulosic impurities such as pectin, protein and fats from the fiber. In addition, energy conservation and environmentally friendly bioscouring process also limit the fiber damage [49].

1.2.6.3 Degumming of plants bast fibers

Pectinase has been used in degumming process of plant fibers such as ramie, sunn hemp, jute, flax and hemp to remove the gum before subjecting them for textile making [31, 46, 50, 51]. The enzymatic degumming process using pectinase with combination of xylanase is environmentally friendly and economic and is excellent replacement of chemical degumming process, which is polluting, toxic and non-biodegradable [46].

1.2.6.4 Retting of plant fibers

In traditional retting process, mixed microbial cultures were used which produced pectinase that releases cellulosic fibers from fiber bundles. As compared to traditional retting, enzymatic retting is faster, controlled and produces fewer odors. Pectinase has been used to separate the fibers from jute and flax by eliminating pectin [49, 51].

1.2.6.5 Wastewater treatment

Typically multiples steps such as physical dewatering, spray irrigation, chemical coagulation and chemical hydrolysis are carried out for the treatment of wastewater from vegetable food industries that mostly contain pertinacious materials. Pectinases used in various industrial processes are a better choice as compared to traditional chemical methods because they are cost effective and ecofriendly. They can eliminate pectin containing substances and can make activated sludge treatment more feasible [33, 49, 52].

1.2.6.6 Coffee and tea fermentation

Pectinase has a significant role in tea and coffee fermentation. In addition to increased tea fermentation, pectinase also destroys the foaming character of instant tea powder by degradation of pectin [53].

1.2.6.7 Paper industry

The uses of different enzymes such as xylanase, ligninase, mannanase, pectinase and α-galactosidase are increasing for biobleaching and papermaking in the paper industry [54, 55]. Pectinase lowers the cationic requirement in papermaking processes [56, 57].

1.2.6.8 Animal feed

Pectinase has been used in feed preparation as multienzyme cocktail with glucanase, xylanase, protease and amylase that reduce the feed viscosity and increase the absorption nutrients, release nutrients either through hydrolysis of non-degradable fibers [49].

2. Conclusion

Pectinase is very important for sustainable industrial processes and can be applied for different industrial applications including food, textile, paper, wine production and research. Pectinase catalyzes the breakdown and modification of pectin-based substances through hydrolysis, trans-elimination and de-esterification reactions. Pectinase can be obtained from different organisms, and microorganisms are usually used for industrial pectinase productions. Microbial pectinases have tremendous applications in different industries and among the top most industrial enzymes that have significance role in the current biotechnological world.

Acknowledgements

The author acknowledged and is thankful to higher education commission Pakistan for financial support by providing local challenge fund to work on enzyme manufacturing.

Conflict of interest

The authors declare no conflict of interest.

Author details

Haneef Ur Rehman
Department of Natural and Basic Sciences, University of Turbat, Turbat, Balochistan, Pakistan

*Address all correspondence to: haneef.baloch@uot.edu.pk

References

[1] Sticklen MB. Plant genetic engineering for biofuel production: Towards affordable cellulosic ethanol. Nature Reviews Genetics. 2008;**9**(6):433-443

[2] Alkorta I, Gabirsu C, Llama MJ, Serra JL. Industrial applications of pectic enzymes: A review. Process Biochemistry. 1998;**33**:21-28

[3] Rombouts FM, Pilnik WL. Pectic enzymes. In: Rose AH, editor. Economic Microbiology: Microbial Enzymes and Bioconversions. Vol. 5. London, England: Academic Press; 1980. pp. 227-282

[4] Vauquelin M. Analyse du tamarin. Annales de Chimie. 1790;**5**:92-106

[5] Braconnot H. Recherches sur un nouvel acide universellement répandu dans tous les végétaux. Annales de Chimie et de Physique. 1825;**28**:173-178

[6] Nussinovitch A. Hydrocolloid Applications: Gum Technology in the Food and Other Industries. Blackie Academic and Professional: London, United Kingdom; 1997. p. 354

[7] Ehrlich F. Die pectin st offe, ihre konstitution und bedeutung. Chemiker Zeitung. 1917;**41**:197-200

[8] Vincken JP, Schols HA, Oomen RJFJ, McCann MC, Ulvskov P, Voragen AGJ, et al. If homogalacturonan were a side chain of rhamnogalacturonan I: Implications for cell wall architecture. Plant Physiology. 2003;**132**:1781-1789

[9] Yapo BM. Pectin Rhamnogalacturonan II: On the “small stem with four branches” in the primary cell walls of plants. International Journal of Carbohydrate Chemistry. 2011. DOI: 10.1155/2011/964521

[10] Zhang J. Biochemical study and technical applications of fungal pectinase. Digital comprehensive summaries of Uppsala dissertations from faculty of science and technology. Acta Universitatis Upsaliensis Uppsala. 2006;**2006**:14

[11] Jayani RS, Saxena S, Gupta R. Microbial pectinolytic enzymes: A review. Process Biochemistry. 2005;**40**:2931-2944

[12] O’Neill MA, Albersheim P, Dravill AG. The pectic polysaccharides of primary cell walls. In: Dey PM, editor. Methods in Plant Biochemistry. Vol. 2. London: Academic Press; 1990. pp. 415-441

[13] Darvill AG, McNeil M, Albersheim P. Structure of plant cell walls VIII. A new pectic polysaccharide. Plant Physiology. 1978;**62**:418-422

[14] Thakur BR, Singh RK, Handa AK, Rao MAD. Chemistry and uses of pectin- a review. Critical Reviews in Food Science and Nutrition. 1997;**37**:47-73

[15] Be Miller JN. An introduction to pectins: Structure and properties. In: Fishman ML, Jem JJ, editors. Chemistry and Functions of Pectins, ACS Symposium Series. Vol. 310. Washington, DC: American Chemical Society; 1986. pp. 2-12

[16] Matsunaga T, Ishii T, Matsumoto S, Higuchi M, Darvill A, Albersheim P, et al. Occurrence of the primary cell wall polysaccharide rhamnogalacturonan II in pteridophytes, lycophytes and bryophytes. Implications for the evolution of vascular plants. Plant Physiology. 2004;**134**:339-351

[17] Voragen AGJ, Coenen G, Verhoef RP, Schols HA. Pectin, a versatile polysaccharide present in plant cell walls. Structural Chemistry. 2009;**20**:263-275

[18] Karr AL. Cell wall biogenesis. In: Bonner J, Varner JE, editors. Plant Biochemistry. New York: Academic Press; 1976. pp. 405-426

[19] Renard CMGC, Thibault JF. Structure and properties of apple and sugar beet pectins extracted by chelating agents. Carbohydrate Research. 1993;**244**:99-114

[20] Hodgson A, Kerr L. Tropical fruit products. In: Walter RH, editor. The Chemistry and Technology of Pectin. San Diego, CA: Academic Press; 1991. pp. 67-86

[21] Fernandez ML. Pectin: Composition, chemistry, physicochemical properties, food applications and physiological effects. In: Cho S, Dreher ML, editors. Handbook of Dietary fiber. Vol. 113. New York: Marcel Dekker, Inc. Food Science and Technology; 2001. pp. 583-601

[22] Mohnen D. Pectin structure and biosynthesis. Current Opinion in Plant Biology. 2008;**11**:266-277

[23] Ceci L, Lozano J. Determination of enzymatic activities of commercial pectinases for the clarification of apple juice. Food Chemistry. 1998;**61**:237-241

[24] Rehman H, Baloch AH, Nawaz MA. Pectinase: Immobilization and applications. A review. Trends in Peptide and Protein Sciences. 2021;**6**:1-16

[25] Pouderoyen GV, Snijder HJ, Benen JAE, Dijkstra BW. Structural insights into the processivity of endopolygalacturonase I from *Aspergillus niger*. FEBS Letters. 2003;**554**:462-466

[26] Yoder MD, Keen NT, Jurnak F. New domain motif: The structure of pectate lyase C, a secreted plant virulence factor. Science. 1993;**260**:1503-1507

[27] Kita N, Boyd CM, Garrett MR, Jurnak F, Keen NT. Differential effect of site-directed mutations in pelC of pectate lyase activity, plant tissue maceration, and elicitor activity. Journal of Biological Chemistry. 1996;**271**:26529-26535

[28] Sharma N, Rathore M, Sharma M. Microbial pectinase: Sources, characterization and applications. Reviews in Environmental Science and Bio/Technology. 2013;**12**:45-60

[29] Maldonado MC, Cáceres S, Galli E, Navarro AR. Regulation of the production of polygalacturonase by *Aspergillus niger*. Folia Microbiologica (Praha). 2002;**47**:409-412

[30] Beg QK, Bhushan B, Kapoor M, Hoondal GS. Production and characterization of thermostable xylanase and pectinase from a *Streptomyces* sp. QG-11-3. Journal of Industrial Microbiology and Biotechnology. 2000;**24**:396-402

[31] Cao J, Zheng L, Chen S. Screening of pectinase producer from alkalophilic bacteria and study on its potential application in degumming of ramie. Enzyme and Microbial Technology. 1992;**14**:1013-1016

[32] Hayashi K, Inoue Y, Shiga M, Sato S, Tokano R, Hirayae K, et al. Pectinolytic enzymes from *Pseudomonas marginalis* maff 03-01173. Phytochemistry. 1997;**45**:1359-1363

[33] Tanabe H, Kobayashi Y, Akamatsu T. Pretreatment of pectic wastewater with pectate lyase from an alkalophilic *Bacillus* sp. Agricultural and Biological Chemistry. 1988;**52**:1855-1856

[34] Schallmey M, Singh A, Ward OP. Developments in the use of *Bacillus* species for industrial production. Canadian Journal of Microbiology. 2004;**50**:1-17

[35] Neves MAD, Kimura T, Shimizu N, Shiiba K. Production of

alcohol by simultaneous saccharification and fermentation of low-grade wheat flour. Brazilian Archives of Biology and Technology. 2006;**49**:481-490

[36] Mojović L, Nikolić S, Rakin M, Vukasinović M. Production of bioethanol from corn meal hydrolyzates. Fuel. 2006;**85**:1750-1755

[37] Hölker U, Lenz J. Solid-state fermentation- are there any biotechnology advantages? Current Opinion in Microbiology. 2005;**8**:301-306

[38] Sunnotel O, Nigam P. Pectinolytic activity of bacteria isolated from soil and two fungal strains during submerged fermentation. World Journal of Microbiology and Biotechnology. 2002;**18**:835-839

[39] Esquivel JCC, Voget CE. Purification and partial characterization of an acidic pectinase polygalacturonase from *aspergillus kawachii*. Journal of Biotechnology. 2004;**110**:21-28

[40] Gummadi SN, Panda T. Purification and biochemical properties of microbial pectinases-a review. Process Biochemistry. 2003;**38**:987-996

[41] Kaur G, Kumar S, Satyanarayana T. Production, characterization and application of a thermostable polygalacturonase of a thermophilic mould *Sporotrichum thermophile apinis*. Bioresource Technology. 2004;**94**:239-243

[42] Gadre RV, Driessche GV, Beeumen JV, Bhat MK. Purification, characterization and mode of action of an endo-polygalacturonase from the psychrophilic fungus *Mucor flavus*. Enzyme and Microbial Technology. 2003;**32**:321-330

[43] Kobayashi T, Higaki N, Suzumatsu A, Sawada K, Hagihara H, Kawai S, et al. Purification and properties of a high molecular-weight, alkaline exopolygalacturonase from a strain of a strain of *Bacillus*. Enzyme Microbiology and Technology. 2001;**29**:70-75

[44] Mohamed SA, Christensen TMIE, Mikkelsen JD. New polygalacturonases from *Trichoderma reesei*: Characterization and their specificities to partially methylated and acetylated pectins. Carbohydrate Research. 2003;**338**:515-524

[45] Mohamed SA, Farid NM, Hossiny EN, Bassuiny RI. Biochemical characterization of an extracellular polygalacturonase from *Trichoderma harzianum*. Journal of Biotechnology. 2006;**127**:54-64

[46] Kapoor M, Beg QK, Bhushan B, Dadhich KS, Hoondal GS. Production and partial purification and characterization of a thermo-alkali stable polygalacturonase from *Bacillus* sp. MG-cp-2. Process Biochemistry. 2000;**36**:467-473

[47] Riou C, Freyssinet G, Fevre M. Purification and characterization of extracellular pectinolytic enzymes produced by *Sclerotiniasclerotiorum*. Applied and Environmental Microbiology. 1992;**58**:578-583

[48] Kashyap DR, Vohra PK, Chopra S, Tewari R. Applications of pectinases in the commercial sector: A review. Bioresource Technology. 2001;**77**:215-227

[49] Hoondal GS, Tiwari RP, Tiwari R, Dahiya N, Beg QK. Microbial alkaline pectinase and their industrial application. Applied Microbiology and Biotechnology. 2002;**59**:409-418

[50] Bruhlmann F, Kim KS, Zimmerman W, Fiechter A. Pectinolytic enzymes from *Actinomycetes* for the degumming of ramie bast fibers. Applied

and Environmental Microbiology. 1994;**60**:2107-2112

[51] Henriksson G, Akin DE, Slomczynski D, Eriksson KEL. Production of highly efficient enzymes for flax retting by *Rhizomucor pusillus*. Journal of Biotechnology. 1999;**68**:115-123

[52] Horikoshi K. Enzymes from alkalophiles. In: Fogarty WM, Kelly CT, editors. Microbial Enzymes and Biotechnology. 2nd ed. Ireland: Elsevier; 1990. pp. 275-295

[53] Carr JG. Tea, coffee and cocoa. In: Wood BJB, editor. Microbiology of Fermented Food. Vol. 2. London: Elsevier; 1985. pp. 133-154

[54] Bajpai P. Application of enzymes in the pulp and paper industry. Biotechnology Progress. 1999;**15**:147-157

[55] Kirk TK, Jefferies TW. Role of microbial enzymes in pulp and paper processing. In: Jefferies TW, Viikari L, editors. Enzymes for Pulp and Paper Processing. ACS Symposium Series. Washington D.C: American Chemical Society; 1996. pp. 1-14

[56] Reid I, Ricard M. Pectinase in papermaking: Solving retention problems in mechanical pulps bleached with hydrogen peroxide. Enzyme and Microbial Technology. 2000;**26**:115-123

[57] Viikari L, Tenkanen M, Suurnäkki A. Biotechnology in the pulp and paper industry. In: Rehm HJ, editor. Biotechnology. 2nd ed. Vol. 10. VCH-Wiley; 2001. pp. 523-546

Chapter 6

Cassava Pectin and Textural Attributes of Cooked gari (*eba*) and fufu Dough

Wasiu Awoyale, Kazeem K. Olatoye and Busie Maziya-Dixon

Abstract

The textural attributes of cooked gari (*eba*) and fufu dough may be affected by the pectin content of the cassava roots; thus, exploring the interaction between pectin and the texture attributes of processed products such as gari and fufu may assist the processors and consumers of the product. The pectin and starch contents, and the composition of the amylose/amylopectin ratio, influence most of the textural changes in roots and tubers during processing, and subsequent preparation for consumption. The textural characteristics of the cooked gari (*eba*) and fufu dough that may be influenced by the pectin content of the cassava roots include hardness, adhesiveness, gumminess, and moldability/cohesiveness. However, there is presently little or no information on the direct relationship between the pectin content of different cassava varieties and the textural attributes of the cooked gari and fufu dough; therefore, there is a need to evaluate the effect of pectin in different cassava varieties on the textural attributes of cooked gari and fufu dough. This will guide gari and fufu producers on the right varieties to be used for gari and fufu to maintain the textural characteristics of the cooked gari and fufu dough preferred by the consumers.

Keywords: pectin, textural attributes, cassava roots, cooked gari (*eba*), fufu dough

1. Introduction

Cassava is widely grown and used in different agroecological zones across many African countries. Cassava utilization patterns vary considerably in different parts of the world and Nigeria; most of the cassava produced (90%) is used for human food. Cassava can be transformed into different products such as gari and fufu [1]. Gari, a roasted solid-state fermented cassava meal, is the most popular product consumed in West Africa and the most important food product in the diet of millions of Nigerians and Ghanaians [1]. Fufu, a submerged fermented and sieved cassava root, is ranked next to gari as an indigenous fermented food in the southern part of Nigeria [2]. Gari and fufu are prepared into the dough by reconstituting in boiled water and consumed with the preferred soup; thus, their textural attributes are key to consumer acceptability of the products [3].

Texture is the sensory and functional manifestation of the quality attributes of foods detected through the senses of vision, hearing, and touch and from kinesthetic qualities [4]. The texture of cooked root and tuber crops and their products are often cited as a primary determinant of the acceptability of improved and local cassava varieties, which may be influenced by higher polysaccharides such as pectin. This is because pectin has been reported to improve the structural and textural attributes of plant-based products. The fundamental constituent of the pectin molecules is D-galacturonic acid monomers in methyl ester conformation linked by α-(1 $\rightarrow$ 4) glycosidic bonds [5]. The biological function of pectin is to cross-link cellulose and hemicellulose fibers, providing rigidity to the cell wall. Pectin is also a major component of the middle lamella, where it helps to bind cells together. It was observed by Franck et al. [6] that cassava roots have different pectin contents depending on the varieties. This was corroborated by Favaro et al. [7], who reported that uronic acid, which is the main constituent of pectin, was extracted from cassava root cell walls. Also, the abundance of pectin structures in cassava cell walls was confirmed by coloring with Coriphosphine-O, which binds to acidic polysaccharides, including pectin [8].

It has been shown that the action of pectin methyl esterase on pectin affects the food textural quality of plant-based food products, either favorably or deleteriously, depending on the product at hand [9]. However, results obtained by Ampe et al. [10] suggest that cell wall degradation is initiated by endogenous pectin esterase located in the intercellular space and released by pH decrease, followed by the action of microbial polygalacturonase and pectate lyase that depolymerizes pectic chains in the cassava roots. Hence, the effect of pectin on the textural attributes of cooked gari and fufu dough may differ because of their different processing methods. Therefore, this chapter aims to discuss the possible effect of pectin on the textural attributes of cooked gari and fufu dough using studies that have been done on pectin and textural attributes of other starchy foods.

2. Production of gari and fufu

Gari is a dry, crispy, creamy-white/yellow and granular product, which is produced by crushing the cassava root into a mash, fermented (lactic fermentation, optional in some locations), dewatered, and sieved into grits. The grits are then roasted manually or mechanically to make the gari [11]. However, the processing of cassava roots into gari differs from one location to another. Some producers/consumers may prefer sour or bland taste gari, fine or coarse particle size gari, palm oil mixed gari, or even gari enriched/fortified with different legumes or protein sources [11–13].

Peeling of freshly harvested cassava roots manually with a knife is most common, but mechanical peelers are now available in countries such as Nigeria and Ghana [12]. The importance of the peeling process is to remove the brown peel, which might affect the gari color and increase its fiber content. Washing of the peeled roots is done to remove all extraneous materials, which could contaminate the gari. Grating of the washed cassava roots is done using a motorized cassava grater, but hand graters, made by fastening the perforated grating sheets on the wood, are still used in some countries. Grating is done to increase the surface area of the cassava root and free up the moisture so that dewatering of the mash can be done easily. The grated cassava mash is bagged using a polypropylene/polyethylene woven bag or basket (lined with

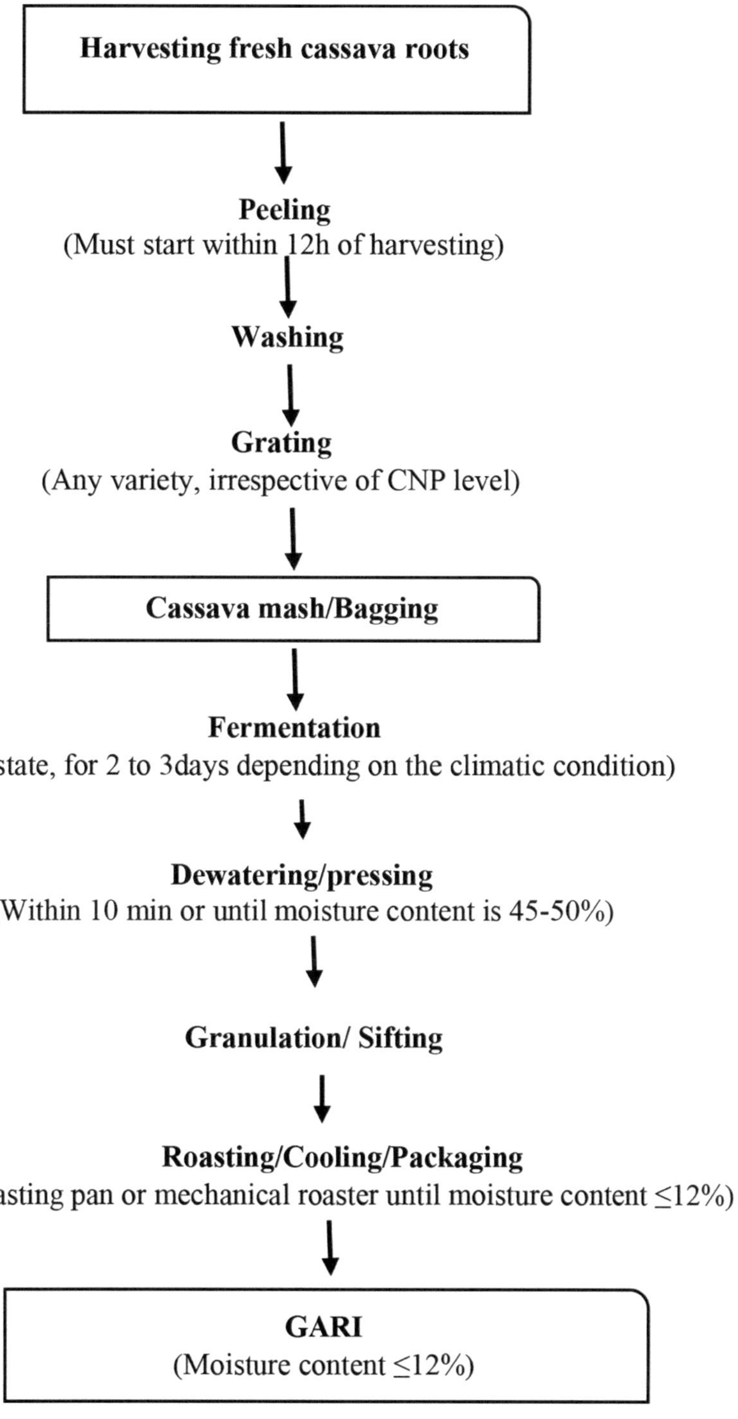

Figure 1.
Production of gari [11].

a polypropylene sack) and left for between 1 and 5 days to ferment, depending on the taste preferred by the consumers. Apart from the taste, fermentation helps to reduce the cyanogenic potential of the product [12].

The fermented mash is then dewatered by pressing with a manual screw or hydraulic press or even wood tied at both ends with a rope, which is still common in most rural communities. Pressing is done to reduce the moisture content of the grated mash before roasting. The cake formed after dewatering is pulverized by a pulverizer/cake breaker or by hand and sieved with a manual woven sieve or rotary sieve, to remove the fiber and lumps. The sieved grit is then roasted, cooled, and packaged (**Figure 1**) [12].

An earthenware stove and a roasting pan made of molded aluminum or stainless steel are used for roasting on a wood fire. In some communities, the roasting pan is smeared with a small amount of palm oil prior to roasting, to produce yellow gari. However, mechanical roasters are now available in Nigeria and Ghana. The roasting

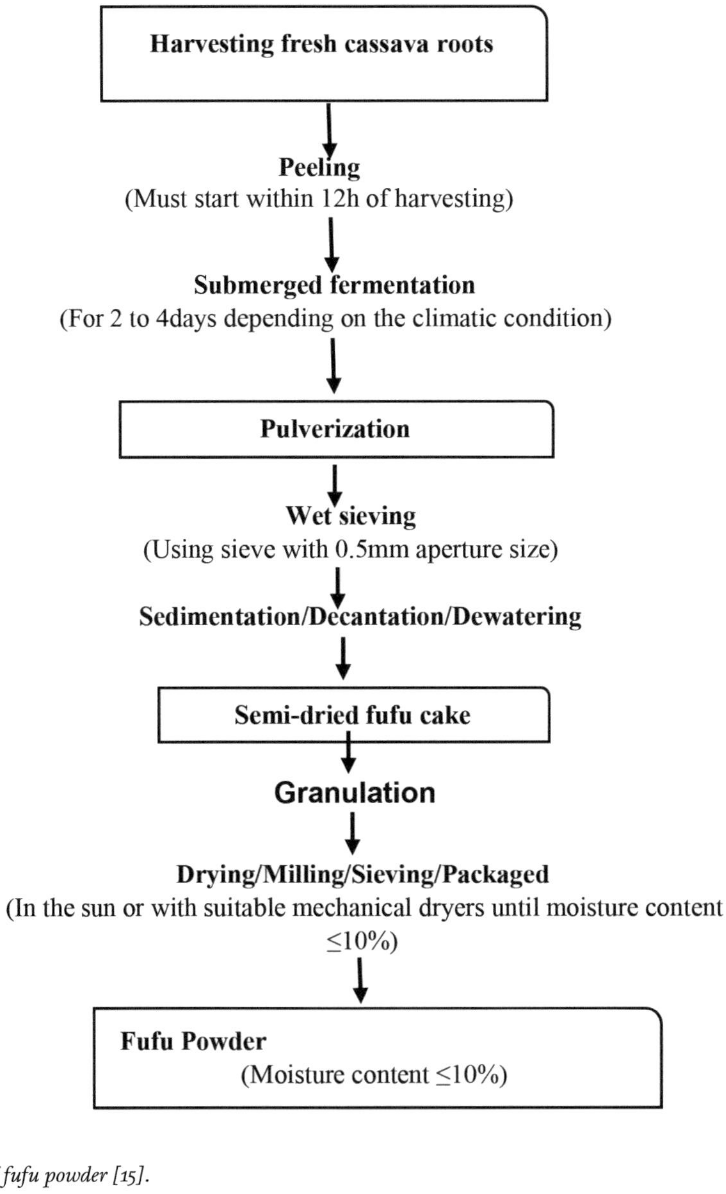

Figure 2.
Production of fufu powder [15].

process develops the gari flavor and improves digestibility, and the extent of drying determines the crispiness and storability of the product. It is important to add that in some communities, the grit is partially toasted and finally dried under the sun, which is not very good as the product will be contaminated. The gari is then cooled for some hours, graded (sieved) depending on the particle sizes preferred by the consumers and packaged depending on the distribution outlet. However, most rural communities package in 50 kg bags for retail. The roasted gari can be consumed in the form of cooked dough (*eba*) with a preferred soup by reconstituting it in boiled water [14]; hence, the textural attributes are very important and may be influenced by the pectin content of the roots.

Fufu is produced by peeling the cassava roots using a stainless-steel knife, washing them with clean water, and soaking them in fermenting drums for four days. The fermented roots are then sieved through a muslin cloth and allowed to form sediment. The sediment is collected and packed in woven polyethylene sacks and dewatered using a manually operated pressing machine. The cake is pulverized and spread on a black polyethylene sheet for drying under the sun. The dried fufu is milled using a hammer mill, cooled, and packaged (**Figure 2**) [15]. Fufu can be sold in a wet form (a semi-solid cake) or in a flour form. Fufu is consumed by reconstituting in boiled water to form a cooked dough, which is consumed with preferred soup [15], thus the need for textural attributes of the fufu that may be affected by the pectin content of the cassava roots.

3. Influence of pectin on the textural attributes of cassava roots

The texture of foods is related to the structure formed by micro- and macromolecular elements forming the cell wall and other regions [16]. Most of the textural changes in roots and tubers during processing are related to pectin and starch contents and the composition of the amylose/amylopectin ratio. Softening of cell walls during the cooking of cassava root was studied for intracellular compounds such as cations (Ca^{2+} and Mg^{2+}), phytic acid, and pectins. It appeared that the cassava variety with the longer cooking time had a lower level of cations and phytic acid and higher levels of chelator-insoluble pectic polysaccharides. It is therefore likely that mealiness is associated with pectins in cassava roots [6]. Maieves et al. [17] reported that cassava varieties whose starch granules are more deeply related to parenchyma tissues, pectin, and cellulose tend to be harder in texture, both in raw and in cooked cassava roots. Infante et al. [18] added that the presence of pectic substances (salts of pectinic and pectic acids, and protopectin) in cassava root may contribute to the texture and hardness of cassava roots, which in turn could be responsible for the mouth feel of cooked or processed foods. A correlation between pectin composition and the cooking quality of boiled cassava roots provided the first evidence that pectins are involved in determining the texture of root and tuber products. In the case of boiled cassava, a soft texture was related to higher levels of methoxylated pectins (i.e., pectins with side groups limiting their ability to form egg-box complexes with Ca^{2+}) and lower levels of nonmethoxylated pectins [19]. This was linked with higher pectin methyl esterase activity, which is associated with increased firmness as the pectin methyl esterase rapidly demethylates pectin in the cell walls and middle lamellae, allowing for hardening through cross-linking with divalent cations [19].

4. Textural attributes of cooked gari (*eba*) and possible influence of pectin

Gari is a roasted, fermented cassava grit, consumed raw, soaked in cold water, or reconstituted in hot water into *eba*, and is common in the diets of millions of people in developing countries [20]. The instrumental texture attributes of *eba* produced from different cassava varieties reported by Awoyale et al. [20] showed that variations exist in their textural attributes. For instance, the hardness of the *eba* was higher in TMS15F1467P0011 gari (54.58 N/m^2) and lower in TMS14F1035P0004 gari (13.71 N/m^2) (**Table 1**). This means that consumers that prefer the firm-textured *eba* can consume the *eba* prepared from the TMS15F1467P0011 gari, while those that prefer the soft-textured *eba* can consume the *eba* prepared from the TMS14F1035P0004 gari. This was in agreement with the hardness of the *eba* prepared from different cassava varieties using the backslopped fermented gari (21.03–30.22 N/m^2) [3]. Hardness is defined as an indicator of the most direct response to taste, which has a direct relationship with chewiness, gumminess, and cohesiveness in the texture profile analysis [3]. It was reported by Zhai et al. [22] that the hardness of the blends of waxy rice starch and pectin gels significantly decreased with the increase in pectin inclusion. The authors attributed this observation to the fact that the pectin formed hydrogen bonds with the waxy rice starch molecules, which interfered with the formation of ordered structures during starch retrogradation. This implies that the pectin content in the TMS14F1035P0004 gari may be high and that of the TMS15F1467P0011 gari may be low, hence the lower hardness of the *eba* prepared from TMS14F1035P0004 gari [22]. This observation was also supported by Gafuma et al. [23]. These researchers reported that pectin contributes to a softer texture of bananas during cooking and cooling, which was attributed to the high water retention capacity of pectins. In addition, the high water binding and holding capacity of pectins may keep the cooked banana structural matrix moist, thus maintaining starch in a gelatinized state, hence making pectin-treated bananas relatively soft in a cooled form [23].

The degree to which the *eba* sticks to the hand, mouth surface, or teeth is known as adhesiveness [3]. The adhesiveness of the *eba* ranged from −177.50 N/m^2 to −39.36 N/m^2, with TMS13F1343P0004 gari having the highest and TMS15F1482P0098 gari the lowest (**Table 1**). This implies that the *eba* prepared from the TMS13F1343P0004 gari may be more adhesive compared to that prepared from the TMS15F1482P0098 gari [3]. Zhai et al. [22] reported that the decrease in adhesiveness in the blends of waxy rice starch and pectin gels with the increase in the pectin content may be associated with the covering of the starch molecules by the pectin, resulting in reduced starch-to-starch hydrophobic interaction. Consequently, the low adhesiveness of the *eba* prepared from TMS15F1482P0098 gari may be attributed to the high pectin content. This is because the lower hydrophobic interaction might have led to inhomogeneity and instability in the network structure of the *eba*, thus reducing the textural characteristics [22].

Usually, the *eba* is squeezed manually, during which the mechanical and geometrical characteristics are assessed, molded into balls with the hand, dipped into the soup, and then swallowed [3]. Hence, moldability is how well the product withstands a second deformation relative to its resistance under the first deformation [3]. The *eba* from TMS13F1053P0010 gari (0.89) had the lowest moldability, and the IITA-TMS-IBA980581 gari (0.98) had the highest moldability (**Table 1**). A similar range of values (0.84–0.98) was reported for the moldability of *eba* prepared from backslopped fermented gari [3]. The high moldability of the *eba* from IITA-TMS-IBA980581 gari may be linked with increased pectin concentration upon cooking and cooling, which may be attributed to the retrogradation of starch [23].

Cassava varieties	Hardness (N/m²)	Adhesiveness (N/m²)	Moldability	Stretchability	Gumminess (N/m²)
TMS13F1343P0004	41.81 ± 0.23hi	39.36 ± 13.60a	0.91 ± 0.00hi	0.96 ± 0.3e–g	38.01 ± 0.20ij
IBA98051	37.64 ± 0.33j	82.31 ± 2.85c–e	0.93 ± 0.00d–h	1.01 ± 0.05a–g	34.96 ± 0.21lm
IBA30572	50.73 ± 1.78bc	83.84 ± 3.54c–e	0.92 ± 0.00f–h	0.97 ± 0.05d–g	46.73 ± 1.54b–d
TMS13F1088P0007	32.00 ± 0.13lm	57.82 ± 1.68ab	0.93 ± 0.02e–h	1.02 ± 0.06a–g	29.66 ± 0.75no
TMS14F1285P0006	31.21 ± 0.76mn	42.67 ± 5.62a	0.93 ± 0.00d–h	0.99 ± 0.01c–g	29.12 ± 0.79o
TMS13F1343P0004	49.58 ± 0.48c	103.22 ± 3.70e–g	0.91 ± 0.04hi	1.04 ± 0.05a–g	45.05 ± 2.64de
TMS13F1160P0004	44.33 ± 1.20ef	98.29 ± 9.11e–g	0.91 ± 0.01hi	1.07 ± 0.09a–f	40.10 ± 0.59gh
TMS13F2110P0008	52.08 ± 0.40b	109.53 ± 25.29f–h	0.91 ± 0.01hi	1.00 ± 0.00b–g	47.05 ± 0.54bc
UYT30106	51.51 ± 2.04b	145.64 ± 12.03ij	0.94 ± 0.01c–h	1.00 ± 0.11c–g	48.07 ± 1.55b
UYT30104	41.19 ± 0.16i	129.69 ± 17.79hi	0.96 ± 0.02a–e	1.07 ± 0.01a–f	39.31 ± 0.54hi
TMS14F1016P0006	30.10 ± 1.24n	68.72 ± 3.94bc	0.95 ± 0.00a–f	1.03 ± 0.04a–g	28.64 ± 1.25op
TMS13F1053P0015	38.29 ± 0.43j	73.97 ± 4.72b–d	0.92 ± 0.02g–i	1.04 ± 0.08a–g	34.94 ± 1.08lm
UYT30111	13.71 ± 0.36p	85.62 ± 2.42c–e	0.95 ± 0.01b–g	1.12 ± 0.04a–c	12.98 ± 0.22q
UYT30112	28.32 ± 0.08o	116.75 ± 4.70gh	0.95 ± 0.00a–f	1.09 ± 0.04a–e	27.00 ± 0.08p
TMS13F1307P0016	41.92 ± 0.06g–i	115.16 ± 0.56gh	0.93 ± 0.01e–h	1.00 ± 0.00b–g	38.60 ± 0.21hi
UYT30109	37.55 ± 0.36j	141.87 ± 1.17ij	0.98 ± 0.00a	1.13 ± 0.07ab	36.81 ± 0.37jk
TMS13F1160P0005	35.61 ± 0.06k	113.29 ± 4.39gh	0.95 ± 0.02b–g	1.07 ± 0.09a–f	33.59 ± 0.71m
TMS13F1153P0001	38.40 ± 0.33j	84.07 ± 10.28c–e	0.93 ± 0.01e–h	1.00 ± 0.04b–g	35.50 ± 0.11kl
TMS13F1049P0001	33.01 ± 0.03l	91.43 ± 13.76d–f	0.92 ± 0.00f–h	1.03 ± 0.00a–g	30.28 ± 0.08no
UYT30108	40.86 ± 1.26i	143.70 ± 3.11ij	0.97 ± 0.02a–c	1.09 ± 0.01a–e	39.35 ± 0.32hi
TMS14F1195P0005	43.43 ± 0.41fg	88.31 ± 7.91c–e	0.92 ± 0.00f–h	1.07 ± 0.01a–f	39.90±0.37gh
TMS14F1285P0017	45.36 ± 0.07e	113.76 ± 2.54gh	0.91 ± 0.00hi	1.06 ± 0.03a–f	41.46 ± 0.09fg
UYT30103	43.41 ± 0.54fg	142.33 ± 0.56ij	0.97 ± 0.01a–c	1.09 ± 0.00a–e	42.04 ± 0.28f
TMEB419 (Control)	37.09 ± 0.01j	118.18 ± 14.42gh	0.95 ± 0.02b–g	1.00 ± 0.00b–g	34.93 ± 0.78lm
UYT30105	47.88 ± 0.62d	158.88 ± 5.35j	0.96 ± 0.00a–d	1.14 ± 0.06a	45.93 ± 0.58cd
TMS14F1035P0004	42.77 ± 0.59f–g	99.40 ± 0.04e–g	0.93 ± 0.00d–h	1.01 ± 0.06a–g	39.83 ± 0.66g–i
UYT30101	54.58 ± 0.08a	145.84 ± 0.47ij	0.96 ± 0.01a–e	1.10 ± 0.01a–d	52.10 ± 0.16a
TMS13F1053P0010	49.56 ± 0.37c	100.50 ± 0.62e–g	0.89 ± 0.01i	0.94 ± 0.04fg	43.92 ± 0.06e
TMS14F1287P0008	34.63 ± 0.01k	83.79 ± 6.15c–e	0.91 ± 0.01hi	0.91 ± 0.10g	31.27 ± 0.12n
UYT30110	45.42 ± 0.40e	177.50 ± 16.45k	0.97 ± 0.00ab	1.06 ± 0.08a–f	43.93 ± 0.51e
Mean	40.46	105.18	0.93	1.04	37.7
p level	*	*	*	**	*

Means with the same letters within the same column are not significantly different ($p < 0.05$);
$^{*}p < 0.05$;
$^{**}p < 0.01$
Values are means of six replicates.
Source: Awoyale et al. [21].

Table 1.
Instrumental texture attributes of eba *produced from different cassava genotypes.*

For consumers that chew *eba* before swallowing, the stretchability is the degree to which the *eba* returns to its original shape after compression between the teeth [3]. The stretchability was higher in the *eba* prepared from the TMS15F1466P0195 gari (1.14) and lower in the *eba* prepared from the TMS14F1287P0008 gari (0.91) (**Table 1**). The high stretchability of the *eba* from the TMS15F1466P0195 gari may be due to the gari's high peak and breakdown viscosities [3]. However, the values of the stretchability of the *eba* prepared from backslopped fermented gari (0.88–1.06) fall within the values of the stretchability of the *eba* in this study [3].

Gumminess is also defined as the energy required to disintegrate a semi-solid food until it can be swallowed [3]. The gumminess of the *eba* ranged from 12.98 to 52.10 N/m^2. *Eba* prepared from TMS14F1035P0004 gari had the lowest gumminess, and the *eba* from TMS15F1467P0011 gari had the highest gumminess (**Table 1**). The gumminess of the *eba* prepared from the backslopped fermented gari (20.54–27.10 N/m^2) falls within the values of the gumminess of the *eba* in this study [3]. The low gumminess of the *eba* prepared from TMS14F1035P0004 gari may be due to the low hydrophobic interaction that might have led to inhomogeneity and instability in the network structure of the *eba*, thus reducing the textural characteristics [22].

5. Textural attributes of cooked fufu and possible influence of pectin

Fufu is a traditional fermented food product consumed in the southern, western, and eastern parts of Nigeria and some other West African countries [24]. The variations in processing methods and differences in the biophysical traits of the varieties may change the textural properties of the cooked fufu [25]. For instance, Awoyale et al. [25] reported that the cooked fufu dough prepared from TMEB419 flour (45.34 N/m^2) was significantly ($p<0.05$) harder than that prepared from TMS13F1153P0001 flour (19.37 N/m^2) (**Table 2**). The high pectin content in the cooked fufu dough prepared from the TMS13F1153P0001 flour might have contributed to the softer texture during cooking and cooling [22]. In addition, the high water binding and holding capacity of pectin may keep the cooked fufu dough matrix moist, thus maintaining starch in a

Samples	Hardness (N/m^2)	Adhesiveness (N/m^2)	Moldability	Stretchability	Gumminess (N/m^2)
NR14B-218	23.43 ± 0.56e	−54.64 ± 8.89b	0.96 ± 0.00c	0.97 ± 0.04ab	22.40 ± 0.61e
TMS13F1153P0001	19.37 ± 1.14f	−37.40 ± 8.24ab	0.99 ± 0.00a	1.02 ± 0.10ab	19.27 ± 1.14f
TMEB419	45.34 ± 1.34a	−43.28 ± 0.90ab	0.93 ± 0.01d	1.06 ± 0.16a	42.38 ± 1.51a
NR1741	26.93 ± 1.16d	−46.44 ± 8.74ab	0.97 ± 0.00b	1.05 ± 0.07a	26.02 ± 1.06d
TMS13F1020P0001	40.10 ± 0.39b	−30.61 ± 11.86a	0.92 ± 0.01e	0.75 ± 0.29b	36.89 ± 0.42b
IITA-TMS-IBA30572	29.45 ± 0.80c	−51.07 ± 13.04b	0.97 ± 0.00b	1.00 ± 0.01ab	28.59 ± 0.84c
Mean	30.77	−43.91	0.96	0.97	29.26
p level	*	NS	*	NS	*

$^{**}p<0.05$.
NS: not significant.
Means with the same letters within the same column are not significantly different ($p > 0.05$).
Source: Awoyale et al. [25].

Table 2.
Instrumental texture profiling of cooked fufu dough produced from different cassava varieties.

gelatinized state, hence making the fufu dough relatively soft in the cooled form [23]. Awoyale et al. [25] added that the hardness of the fufu dough has a positive correlation with most of the functional properties of the flour (except for dispersibility) and a negative correlation with pasting properties (except for breakdown viscosity and pasting temperature). Although this observation was not supported by the findings of Zhai et al. [22] who stated that as pectin content increased in the waxy rice starch and pectin blends, the peak and final viscosity of the products gradually increased. This may be due to differences in their starch and pectin composition.

Adhesiveness is the degree to which the cooked dough sticks to the hand, mouth surface, or teeth [25]. The adhesiveness of the fufu dough ranged from −54.64 to −30.61 N/m^2, with the product from TMS13F1020P0001 flour having the highest value ($p<0.05$) and that from NR14B-218 flour having the lowest (**Table 2**). The low adhesiveness of the cooked fufu from NR14B-218 may be attributed to the lower hydrophobic interaction that may have led to inhomogeneity and instability in the network structure of the fufu dough, thus reducing the textural characteristics [22]. The adhesiveness of the cooked fufu dough had a positive correlation with all the functional properties (except for the water absorption capacity and solubility index), pasting properties, and chemical composition (except for sugar and starch, ash content, and pH value) of the flour [25]. Since high water-holding capacity is a characteristic of pectins, the negative correlation that exists between the adhesiveness of the cooked fufu dough and the water absorption capacity of the fufu flour might be a sign that the fufu flour from the NR14B-218 variety may have low pectin content [16].

Cohesiveness and moldability define how well the cooked fufu dough withstands a second deformation relative to its resistance to the first time. It is calculated as the work area during the second compression divided by the work area during the first [26]. Usually, the cooked fufu dough is squeezed manually, during which the mechanical and geometrical characteristics are assessed, molded into balls with the hand, then dipped into the soup, and swallowed [25]. The moldability of the fufu dough ranged from 0.92 in TMS13F1020P0001 flour to 0.99 in TMS13F1153P0001 flour (**Table 2**). The high cohesiveness of the cooked fufu dough from TMS13F1153P0001 flour may be linked with increased pectin concentration upon cooking and cooling, which may be due to the retrogradation of starch [16]. The moldability of the fufu dough was positively correlated with all the functional properties of the flour (except for water-absorption capacity, swelling power, and dispersibility), the pasting properties (except for peak and breakdown viscosities and pasting temperature), and the chemical composition (except for amylose content) [25]. The positive correlation between the cooked fufu dough moldability and the final viscosity may be evidence that the TMS13F1153P0001 fufu flour is high in pectin content [22]. Also, the possible interaction between the starch molecules and pectin during the gelatinization process may have increased the final viscosity of the dough and then responsible for the high moldability of the cooked fufu dough from TMS13F1153P0001 flour [27].

Stretchability or elasticity is the degree to which the cooked fufu dough returns to its original shape after compression between the teeth [25, 26]. The stretchability was lower in the product from TMS13F1020P0001 flour (0.75) and higher in that from TMEB419 flour (1.06) (**Table 2**). The stretchability of the dough also had a positive correlation with all the functional properties of the flour (except for solubility index and dispersibility), pasting properties (except for setback viscosity, peak time, and pasting temperature), and chemical composition (except for starch and amylose contents) [25].

The energy required to disintegrate a semi-solid food until it can be swallowed is known as gumminess. It is calculated as cohesiveness multiplied by hardness [25, 26]. The gumminess of the product from TMEB419 flour (42.38 N/m^2) was significantly ($p<0.05$) more than that in the product from TMS13F1153P0001 (19.27 N/m^2) (**Table 2**). The low gumminess of the cooked fufu dough prepared from TMEB419 flour may be due to the low hydrophobic interaction that might have led to inhomogeneity and instability in the network structure of the dough, thus reducing the textural characteristics [22]. Awoyale et al. [25] added that the gumminess of the fufu dough has a positive correlation with the functional properties of the flour (except for dispersibility), a negative correlation with the pasting properties (except for breakdown viscosity and pasting temperature), and a negative correlation with the chemical composition (except for amylose content). The positive correlation between the gumminess of the fufu dough and the breakdown viscosity implies that the TMS13F1153P0001 cooked fufu dough with low gumminess may be high pectin content. This is because Luo et al. [27] reported that the addition of pectin gradually decreased the breakdown value of the waxy rice starch and pectin blends. The authors added that the decrease in breakdown values might be due to the pectin being able to cover the starch granules step by step with the increasing concentration of pectin so that the stability of the waxy rice starch and pectin mixture was enhanced. Also, the negative correlation between the gumminess of the fufu dough and the setback viscosity of the fufu flour may be evidence that TMS13F1153P0001 cooked fufu dough with low gumminess maybe high in pectin content. This is because Zhai et al. [22] reported that the addition of pectin significantly reduced the setback viscosity of the waxy rice starch and pectin mixture, implying that pectin inhibited the retrogradation of gelatinized starch.

6. Conclusions

The pectin and starch contents, and the composition of the amylose/amylopectin ratio, influence most of the textural changes in roots and tubers during processing and subsequent preparation for consumption. The textural characteristics of the cooked gari (*eba*) and fufu dough that may be influenced by the pectin content of the cassava roots include hardness, adhesiveness, gumminess, and moldability/cohesiveness. However, there is presently little or no information on the direct relationship between the pectin content of different cassava varieties and the textural attributes of the cooked gari and fufu dough; therefore, there is a need to evaluate the effect of pectin in different cassava varieties on the textural attributes of cooked gari and fufu dough. This will guide gari and fufu producers on the right varieties to be used for gari and fufu to maintain the textural characteristics of the cooked gari and fufu dough preferred by the consumers.

Conflict of interest

The authors declare no conflict of interest.

Author details

Wasiu Awoyale[1,2], Kazeem K. Olatoye[2] and Busie Maziya-Dixon[1*]

1 Food and Nutrition Sciences Laboratory, International Institute of Tropical Agriculture, Ibadan, Oyo State, Nigeria

2 Department of Food Science and Technology, Kwara State University Malete, Kwara State, Nigeria

*Address all correspondence to: b.maziya-dixon@cgiar.org

References

[1] Afoakwa EO, Adjonu R, Asomaning J. Viscoelastic properties and pasting characteristics of fermented maize: Influence of the addition of malted cereals. International Journal of Food Science and Technology. 2010;**45**:380-386

[2] Chijioke U, Madu T, Okoye B, Ogunka AP, Ejechi M, Ofoeze M, et al. Quality attributes of fufu in South-East Nigeria: A guide for cassava breeders. International Journal of Food Science and Technology. 2021;**56**(3):1247-1257

[3] Awoyale W, Oyedele H, Adenitan AA, Adesokan M, Alamu EO, Maziya-Dixon B. Relationship between quality attributes of backslopped fermented gari and the sensory and instrumental texture profile of the cooked dough (*eba*). Journal of Food Processing and Preservation. 2022;**2022**:e16115. DOI: 10.1111/jfpp.1611

[4] Civille GV, Ofteda KN. Sensory evaluation techniques — Make "good for you" taste "good". Physiology & Behavior. 5 Nov 2012;**107**(4):598-605. DOI: 10.1016/j.physbeh.2012.04.015

[5] Wang S, Luo H, Zhang J, Zhang Y, He Z, Wang S. Alkali-induced changes in functional properties and in vitro digestibility of wheat starch: The role of surface proteins and lipids. Journal of Agricultural and Food Chemistry. 2014;**62**(16):3636-3643

[6] Franck H, Christian M, Noel A, Brigitte P, Joseph H, Cornet D, et al. Effects of cultivar and harvesting conditions (age, season) on the texture and taste of boiled cassava roots. Food Chemistry. 2011;**126**(1):127-133

[7] Favaro SP, Beleia A, da Silva Fonseca N, Waldron KW. The roles of cell wall polymers and intracellular components in the thermal softening of cassava roots. Food Chemistry. 2008;**108**:220-227. DOI: 10.1016/j.foodchem.2007.10.070

[8] Staack L, Della Pia EA, Jørgensen B, Pettersson D, Rangel N. PedersenCassava cell wall characterization and degradation by a multicomponent NSP-targeting enzyme (NSPase). Scientific Reports. 2019;**9**:10150. DOI: 10.1038/s41598-019-46341-2

[9] Jolie RP, Duvetter T, Van Loey AM, Hendrickx ME. Pectin methylesterase and its proteinaceous inhibitor: A review. Carbohydrate Research. 2010;**345**:2583-2595. DOI: 10.1016/j.carres.2010.10.002

[10] Ampe F, Keleke S, Robert H, Brauman A (1995). The role of pectin degrading enzymes during cassava retting. In Tagbor Egbe, Abrauman, D. Griffon, S. Trèche (éd), Transformation Alimentaire du manioc. Paris, ORSTM

[11] Awoyale W. Training Manual for the Production of Cassava Products in Liberia. Ibadan, Nigeria: International Institute of Tropical Agriculture (IITA); 2018. p. 44

[12] Abass AB, Dziedzoave NT, Alenkhe BE, James BD. Quality Management Manual for the Production of gari. Ibadan: International Institute of Tropical Agriculture; 2012. p. 50

[13] Olaleye OO, Afolayan SS, Aondo TO, Adeniyi AB. Effect of storage methods on the physicochemical and sensory attributes of 'gari' processed from stored cassava root varieties. Journal of Agriculture and Ecology Research International. 2018;**14**(4):1-10

[14] Udoro EO, Kehinde AT, Olasunkanmi SG, Charles TA. Studies on the physicochemical, functional and sensory properties of gari processed from dried cassava chips. Journal of Food Processing & Technology. 2014;**5**:1-8

[15] Awoyale W. Hazard Analysis and Critical Control Point (HACCP) System In the Production Of High-Quality Cassava Flour, Fufu Powder and Gari in Liberia. Ibadan, Nigeria: International Institute of Tropical Agriculture (IITA); 2018. p. 4

[16] Gafuma S, Byarugaba-Bazirake G, Mugampoza E. Textural hardness of selected Ugandan banana cultivars under different processing treatments. Journal of Food Research. 2018;**7**(5):98-111 DOI: 10.5539/jfr.v7n5p98

[17] Maieves HP, Oliveira DC, Bernardo C, Müller CMO, Amante ER. Microscopy and texture of raw and cooked cassava (Manihot esculenta Crantz) roots. Journal of Texture Studies. 2011;**2**:1-10

[18] Infante RB, García OO, Rivera C. Characterization of dietary fiber and pectin of cassava bread obtained from different regions of Venezuela. Revista Chilena de Nutricion. 2013;**40**(2):169-173

[19] Maziya-Dixon B, Mestres C, Bugaud C, Dahdouh L, Tran T. Biophysical Characterization of Quality Traits – RTBfoods Scientific Progress Report for Period 3 (Jan-Dec 2020). Ibadan, Nigeria: RTBfoods Scientific Progress Report; 2021. p. 89

[20] Abass AB, Awoyale W, Ogundapo A, Oluwasoga O, Nwaoliwe G, Oyelekan J, et al. Adoption of improved cassava varieties by processors is linked to processing characteristics and products' biophysical attributes. Journal of Food Processing and Preservation. 2022;**00**:e16350. DOI: 10.1111/jfpp.16350

[21] Awoyale W, Oyedele H, Adesokan M, Alamu EO, Maziya-Dixon B. Can improved cassava genotypes from the breeding program substitute the adopted variety for gari production? Biophysical and textural attributes approach. Frontiers in Sustainable Food Systems. 2022;**6**:984687. DOI: 10.3389/fsufs.2022.984687

[22] Zhai Y, Xing J, Luo X, Zhang H, Yang K, Shao X, et al. Effects of pectin on the physicochemical properties and freeze-thaw stability of waxy rice starch. Food. 2021;**10**(10):2419. DOI: 10.3390/foods10102419

[23] Armstrong E, Eastwood M, Brydon W. The influence of wheat bran and pectin on the distribution of water in rat caecal contents and faeces. British Journal of Nutrition. 1993;**69**(3):913-920. DOI: 10.1079/BJN19930091

[24] Rosales-soto MU, Ross CF, Younce F, Fellman JK, Mattinson DS, Huber K, et al. Physicochemical and sensory evaluation of cooked, fermented protein-fortified cassava (Manihot esculenta Crantz) flour. Advances in Food Technology and Nutritional Sciences Open Journal. 2016;**2**:9-18

[25] Awoyale W, Oyedele H, Adenitan AA, Adesokan M, Alamu EO, Maziya-Dixon B. Correlation of the quality attributes of fufu flour and the sensory and instrumental texture profiles of the cooked dough produced from different cassava varieties. International Journal of Food Properties. 2022;**25**(1):326-343. DOI: 10.1080/10942912.2022.2026955

[26] Goddard J, Harris KP, Kelly A, Cullen A, Reynolds T, Anderson L. Root,

Tuber, and Banana Textural Traits. A Review of the Available Food Science and Consumer Preferences Literature. Prepared for the Agricultural Policy Team of the Bill & Melinda Gates Foundation by the Evan School Policy Analysis and Research (EPAR). 2015, p. 52.

[27] Luo S, Chen R, Huang L, Liang R, Liu C, Chen J. Investigation on the influence of pectin structures on the pasting properties of rice starch by multiple regression. Food Hydrocolloids. 2017;**63**:580-584